Machinery Lubrication and Reliability

Machinery Lubrication and Reliability

Dr. Trinath Sahoo

INDUSTRIAL PRESS, INC.

Industrial Press, Inc.

32 Haviland Street, Suite 3
South Norwalk, Connecticut 06854
Phone: 203-956-5593
Toll-Free in USA: 888-528-7852
Email: info@industrialpress.com

Author: Dr. Trinath Sahoo
Title: Machinery Lubrication and Reliability
Library of Congress Control Number: 2019951915

© by Industrial Press, Inc.
All rights reserved. Published in 2020.
Printed in the United States of America.

ISBN (print):	978-0-8311-3638-3
ISBN (ePUB):	978-0-8311-9507-6
ISBN (eMOBI):	978-0-8311-9508-3
ISBN (ePDF):	978-0-8311-9506-9

Publisher/Editorial Director: Judy Bass
Copy Editor: James Madru
Compositor: Patricia Wallenburg, TypeWriting
Proofreader: Claire Splan
Indexer: Karin Arrigoni

books.industrialpress.com
ebooks.industrialpress.com

Contents

Preface

Companies who want to be globally competitive must try to find the optimum balance in both quality and cost of manufacture in order to be the most competitive producer. Equipment reliability is a key component in overall competitiveness. Maintenance organizations have a tremendous impact in achieving high reliability of the plant's equipment to improve quality and lower operational costs. The cost of equipment downtime is normally higher than the cost of a well-designed and maintained piece of equipment. A firm foundation in fundamental maintenance practices helps enhance reliability of plant and equipment. To get maximum benefit of the advanced maintenance operational strategies, an excellent understanding of equipment lubrication is a prerequisite.

The correct lubricant properties for the application, contamination control and analysis, and lubrication handling process are imperative to achieve the best life-cycle costs for equipment. Being proactive in determining lubrication problems and concerns is a much less expensive alternative than allowing excessive friction to ruin a piece of equipment.

This book offers engineers a clear interdisciplinary introduction and orientation to all major aspects of lubricant selection and application. The author provides guidance to lubrication practice in industry, with the emphasis on practical application. He covers all aspects of a lubrication program starting with how lubrication reduces friction and wear, properties, and selection of lubricants for a wide range of uses and the factors that determine their suitability, lubricant application for different industrial machinery, factors to be considered in storage and handling, correct sampling procedures, oil analysis to predict equipment problems as well as deterioration of lubricants, and finally how to create an effective lubrication program development.

The author worked as a maintenance and reliability head for more than 25 years in different process plants (power, petroleum, refinery, and petrochemical) and tries to share his experience and wisdom in the book. This book took many years to complete as many problems encountered at sites and their practical solutions are narrated in a clear way.

For those looking to obtain an ICML certificate, there are questions and answers at the end of each chapter. And rounding out this great package are additional questions on the affiliated web site at www.machinerylubricationreliability.com.

This book is intended to help engineers in industry who work in the field of operation and maintenance of machinery. It gives information that these engineers need in a form that is instantly accessible and easy to read.

Acknowledgments

First and foremost, I would like to thank God, the Almighty, for His showers of blessings throughout to complete this book successfully. In the process of putting this book together, I realized how true this gift of writing is for me. You've given me the power to believe in my passion and pursue my dreams. I could never have done this without the faith I have in you, the Almighty.

To my parents, I am speechless! I can barely find the words to express all the wisdom, love, and support you've given me. You were my number one fan and for that I am eternally grateful. If I am blessed to live long enough, I hope I will be as good a father to my children as you are and always have been to me.

To my wife, Chinoo: What can I say? You are one of the main reasons that it was great! I am so thankful that I have you in my corner pushing me when I am ready to give up. All the good that comes from this book I look forward to sharing with you! Thanks for not just believing, but knowing that I could do this! I love you always and forever!

To my children Sonu and Soha: You are the most prized possession that I have ever gotten in my life! The hard work that went into making this book a reality is all because of you! You gave me something to work for. A better life for you than I had is all I want. I love you more than you will ever know and my book is proof of the beauty I see whenever I look into your eyes!

Also thanks to editors, publishers, and other people who helped bring this book to life. Love you all!

Machinery Lubrication and Reliability

CHAPTER 1

Introduction

LINK BETWEEN MAINTENANCE STRATEGY AND LUBRICATION RELIABILITY

Maintenance costs are a major part of the total operating costs in all manufacturing plants. Depending on the type of industry, maintenance costs can represent between 15 and 60 percent of the cost of goods produced. Hence reliability of plant and equipment has a significant impact on profits in any organization. Reliability can be achieved by choosing appropriate maintenance strategies to generate continuous operation of the plant's equipment. Recent surveys of maintenance effectiveness have shown that one-third of maintenance costs are wasted because of unnecessary or improperly carried out maintenance. Waste reduction in maintenance impacts cost reduction, profit maximization, and reduction of negative environmental impacts. In addition, equipment reliability and availability are major factors in the implementation of lean manufacturing. Avoiding unnecessary maintenance of equipment requires a shift from the traditional repair-focused maintenance culture to a lifecycle-optimized, proactive, condition-directed maintenance management system. Fundamental maintenance practices such as reactive, preventive, predictive, and proactive maintenance are the key enablers for an efficient advanced maintenance management system.

To get maximum benefit from advanced maintenance strategies, an excellent understanding of equipment lubrication is required. The applied maintenance strategies are not effective as stand-alone initiatives unless basic asset care and lubrication needs are integrated into those strategies. To achieve the highest return on maintenance investment, a reliability-centered lubrication (RCL) strategy should be incorporated into the existing maintenance strategy. In this chapter we will discuss how asset maintenance strategy and lubrication reliability are interlinked.

ASSET MAINTENANCE STRATEGY

As industry has evolved, so have the philosophies and practices of maintenance and lubrication. Maintenance managers often wait until a component fails before they take action to repair or replace it. In many organizations, managers do periodic maintenance on equipment to keep it running efficiently. Finding the appropriate balance of maintenance approaches is the key to minimizing asset downtime and repair costs while maintaining a safe environment for workers.

Four basic types of asset maintenance strategies have been developed over the last few decades:

- Reactive
- Preventive
- Predictive
- Proactive

Reactive Maintenance

Reactive maintenance, also known as the *run-to-failure strategy,* occurs when you take action after an asset fails. Because you only spend money when something breaks, the reactive maintenance approach might seem cheaper, but it costs you more in the long run. Reactive maintenance shortens the life of assets and may cause them to break down more frequently. Too often reactive philosophies are adopted by organizations that are short on personnel or believe in constant "firefighting." This leads to overblown maintenance budgets and poor operational performance. This maintenance philosophy is not sustainable and has largely been relegated to noncritical or small pieces of equipment. When an organization adopts reactive maintenance mode, its daily maintenance activities are often driven by unforeseen problems.

Preventive Maintenance

Preventive maintenance occurs when maintenance takes place before something breaks down. It is a time-based approach that is carried out at predetermined intervals to reduce failure risk or performance degradation of assets. In cases where safety is paramount, planned or scheduled maintenance is implemented to move away from the reactive state. This type of maintenance is scheduled based on the number of operating hours or calendar time. This includes closely following original equipment manufacturer (OEM) recommendations or intervals to prevent a failure. The aim of preventive maintenance is to minimize unplanned downtime and reduce repair costs. Although preventive maintenance can help reduce the chaos of failures, it can still result in high maintenance costs when good parts are replaced.

Predictive Maintenance

Recently, new tools and accessories have become available to aid in equipment monitoring and catching potential issues earlier. This monitoring of failure symptoms and faults is known as *predictive maintenance*. Predictive maintenance is a condition-based approach to maintenance. Rather than servicing assets on a fixed schedule, one can evaluate the condition of components to determine whether they need to be serviced. Examples of predictive maintenance include oil analysis, thermal analysis, vibration analysis, ultrasound, thermography, and a host of other technologies to provide an early warning of an impending problem. Predictive maintenance works well for machines that run continuously and often results in a reduction of unplanned downtimes. However, it usually comes with considerable upfront costs, not just for the necessary tools but also for training the individuals who are expected to capture the pertinent data. Diligence is required to ensure that data are collected from the same place and in the same manner each time. Inconsistent practices will skew the data and make it much more difficult to take appropriate action.

Proactive Maintenance

As a condition-based maintenance program matures, maintenance managers with experience can make continuous improvements to maintenance activities by the use of *proactive maintenance* (PAM). Rather than fixing machines, proactive maintenance eliminates what causes them to fail. It is a concept based on "learning from experience" in maintenance work. This approach involves the use of direct feedback from maintenance personnel and findings from preventive maintenance checks, failure causes, and equipment monitoring to improve the effectiveness of the maintenance work. The proactive approach responds primarily to equipment assessment and root-cause analysis, making appropriate adjustments to the maintenance task to eliminate deficiencies in the future. It can be used to extend equipment life, as opposed to simply improving the process for repairs or identifying when a machine is going to fail. Without a proactive mind-set, equipment failures will continue to plague most maintenance departments. Analyzing what went wrong and taking steps to prevent it from happening again are the focus of being proactive.

RELIABILITY-CENTERED MAINTENANCE

Reliability-centered maintenance (RCM) is an industrial maintenance technique based on the analysis of system functions, consequences, and failure modes of process assemblies or components. This method consists of seeking the most cost-effective maintenance technique while limiting the risk of failure and providing an optimal context for maintenance technicians. Easy to implement, RCM differs from current practices (as described in the preceding section) because it is essentially based on commonsense and organization; it results in the creation of a project group involving different departments and the use of well-known maintenance analysis tools

RCM involves the use of various tools that are well known to maintenance professionals, including failure modes, effects, and criticality analysis (FMECA) and the decision tree. FMECA templates can be developed at a class/subclass/qualifier level (e.g., pump/centrifugal/coupled or pump/centrifugal/belt driven). Significant time savings can be realized by developing templates

Once the FMECAs are completed, they can be applied at the asset level. This more granular review ties in the criticality ranking criteria to determine whether the consequences of failure are great enough to perform the task. This is really an economic decision rule: "Is the cost of failure greater than the cost to mitigate?" This is extremely important to note because the goal of these programs is to reduce the cost of maintenance while maintaining high asset utilization. FMECA also uses other tools, such as criticality matrices, different process validation techniques, and a solid spare parts management strategy.

RCM addresses the cause of failures and the consequence of failures and asks whether the consequence matters. If it does not, and this is the case for many failure modes, breakdown maintenance or run-to-failure strategies make economic sense. One of the things researchers found is that a large proportion of failure modes are not age related, so doing age-related maintenance (preventive maintenance in common language) does not make technical sense. For such failures, predictive maintenance is the preferred solution. RCM also helps us define how often we should take the readings. Preventive maintenance is applicable for only a limited number of failure modes; the trick is to find the right ones. RCM show us how to do so, as well as how often. There is an important class of failures known as *hidden failures*. For these we need detective strategies (predictive maintenance). Again, RCM show us how and how often.

DEVELOPING THE RIGHT MAINTENANCE STRATEGY FOR YOUR ASSETS

Many companies have recently implemented reliability initiatives geared toward optimizing the maintenance function at their plants. There are many approaches to successfully implementing a reliability program and maintenance strategy. Let's discuss a proven model for improving a company's reliability-based maintenance program. Assign a maintenance strategy to each asset you have based on its level of criticality. Assets that have a high consequence of failure are considered highly critical assets. Monitor the condition of highly critical assets continuously with a predictive maintenance plan to protect them and predict failures. Assets of low to mid-level criticality should be monitored with preventive maintenance. A run-to-failure strategy (reactive maintenance) can be used for assets that aren't considered essential.

Reactive maintenance should only be used if the consequence of failure is so low that it makes sense to allow the asset to fail rather than spend valuable maintenance time performing predictive or preventive maintenance tasks. Most companies find that they have to use a combination of predictive, preventive, and reactive maintenance strategies for the best results.

WHAT IS LUBRICATION RELIABILITY?

Investigations conducted on why bearings fail reveal the alarming fact that over 60 percent of the damage is lubrication related. Lubrication practices within a plant have a direct effect on plant and equipment reliability. When the lubrication is working effectively, wear will be reduced, and equipment reliability will be improved. A lubrication reliability strategy focuses on all parameters that protect the average lubrication film thickness, thereby reducing component wear and increasing equipment reliability. Like reliability-centered maintenance, *reliability-centered lubrication* (RCL) is a logical way to identify which equipment needs to be maintained on a condition-monitoring and preventive maintenance basis rather than a *run-to-failure* (RTF) maintenance strategy. Thus, when the lubrication works in a reliable way, the equipment reliability will improve, meaning that a lubrication reliability strategy is all about ensuring that effective machine lubrication occurs within the machine, resulting in reduced wear and failures.

The RCL program optimizes lubrication maintenance practices by:

- Designing the applied tasks based on engineered applications specific to component attributes and operational and environmental considerations
- Standardizing component modifications to ease the burden of completing basic lubrication tasks
- Maximizing the required task intervals for a component
- Optimizing the number of correct lubricants on site
- Quantifying component health based on observational and trended data

MAINTENANCE STRATEGY APPLICABLE TO LUBRICATION RELIABILITY

All four of the preceding maintenance strategies can be applied to oil analysis and lubrication reliability:

Reactive lubrication. A reactive lubrication strategy occurs when an oil or grease sample is taken only after a potential problem is identified via a sensory inspection. Oil is topped up based on an abnormal inspection result, such as a sight glass showing a low oil level or a chain that appears dry. In many machines, an oil level that is too low can have catastrophic effects.

Action must be taken immediately in these cases to ensure that no lasting damage occurs. Similarly, when performing grease lubrication in a reactive state, you wait until an issue is apparent before adding grease to a bearing. But greasing in response to a noise or elevated temperature is very reactive. By the time these symptoms arise, damage has already occurred.

Preventive lubrication. In this type of lubrication strategy, routine samples are extracted, but the results are not analyzed. Changing the oil based on a time period or operational interval is common for most noncritical or small-volume machines, but it can lead to replacing oil that is still good or going far too long between oil changes. Greasing a machine according to a calendar date is pervasive in the industry, but adding grease based on time may lead to overgreasing or undergreasing the machine. This can be wasteful in terms of both personnel and lubricant.

Predictive lubrication. In this type of lubrication strategy, good samples are obtained and analyzed, and action is taken based on results from the laboratory. Using oil analysis to identify the proper oil change interval is the best approach for large oil volumes and critical machines. When an oil sample is tested, you can distinguish many of its characteristics and determine whether it should remain in service and how much more life it may have. This greatly improves your decision-making ability and can minimize the impact of a lubricant failure by planning for a shutdown or switching to an auxiliary machine.

Proactive lubrication. In this type of lubrication strategy, new lubricants are sampled prior to service. Samples are taken from the right place, in the right way, using the right tests and with the right interpretation strategy. To be proactive when oiling a machine, you must eliminate the root causes of failure. This is accomplished by ensuring that the proper oil is applied and that it is clean and defect free. Your storage and handling practices should be examined and improved to make certain that lubricants are as clean as possible when they reach the machine. This includes filtering the oil prior to service and using transfer containers that can be hermetically sealed. These practices will reduce the number of failures experienced at your plant.

MANAGING A SUCCESSFUL LUBRICATION PROGRAM

Many organizations struggle when implementing a lubrication program because they have only a partial vision of the program's scope. However, effective program administration with a systematic point of view can help you achieve the goal of lubrication excellence. Successful implementations of lubrication best practices consider several technical, organizational, and human factors related to a lubrication project. These principles are suitable not only for lubrication programs but also for other maintenance strategies. A good lubrication program can be implemented by applying the following steps:

Steps to Reliability-Centered Lubrication

1. **Lubrication Assessment.** The lubrication assessment is used to help a maintenance manager understand where he or she stands on lubrication practices. Precision lubrication is about far more than what oil or grease you use. It needs to focus on all aspects of the lubrication management process, starting from selecting the right lubricant, storage and handling, dispensing, and contamination control. And for assets that require oil analysis, getting representative oil samples that are reliable is critical in making the right maintenance decisions. A lubrication assessment tool measures your current lubrication procedures against industry best practices, highlighting the strengths and weaknesses of your current lubrication program across key areas, thus helping you identify opportunities for improvement. Once this is done, an action plan must be developed to implement changes in the system to transform the culture.

2. **Organization and Planning.** To start a lubrication program, it should be centralized to obtain the benefits gained by rationalizing and then standardizing lubrication-related processes. The four main stakeholders of a lubrication program will be the maintenance department (customer), the planning department (which will distribute the schedules), the lubrication services team (who will execute the tasks), and the lubrication subject matter experts (who must provide up-to-date recommendations and best practices regarding plant lubrication needs). These experts also could include plant lubrication engineers, lubricant suppliers, OEMs, and/or other lubrication engineering subject matter experts.

   It is advised that a central department, such as a planning cell, should lead the process and manage the functions once the program is in place. This is the best scenario because the program needs to be implemented throughout the entire plant.

3. **Identification and Inspection.** Operators and maintenance technicians regularly walk-down machines and looked them over every day. Taking this opportunity, they must verify whether machinery lubrication is acceptable or not. Regular inspections of oil levels (min-max level), water separation through a sight-glass inspection and tap, possible lubricant and/or grease leakage, blocked grease lines, grease levels in automatic lubricators, and storage inspection (open drums, dirty connectors, lost protection caps) are also part of a reliable lubrication management program .

4. **Lubricant Storing and Dispensing.** The proper storing, handling, and dispensing of lubricants are the first steps in helping to protect plant personnel against

health hazards and minimize the risk of environmental contamination. They also help in reducing contamination of the oil and grease

5. **Avoid Oil Mix-up.** Oil mix-up is one of the most common lubrication problems that can affect machinery reliability. Putting the right lubricating oil in the right equipment is one of the simplest tasks to improve machine reliability. Lubricants are normally formulated with a balance of performance additives and base stocks to match the lubrication requirements. When lubricants are mixed, this balance becomes upset. To reduce the chances of oil mix-up, the viscosity, brand name, and grade of new oil should be checked.

6. **Cleanliness Control.** In many industries, because of the process materials and ambient conditions, cleanliness is difficult to maintain, although cleanliness starts with a clean operating environment, and plant managers should focus on this aspect. Machine cleanliness and general housekeeping are commonly overlooked, and as a result, contamination control programs become ineffective.

7. **Lubricant Sampling.** How accurately the oil samples are representative of the oil in use plays a key role in wear particle analysis. Special consideration must be paid to the location, sampling method, and frequency of testing, but simple adherence to the given recommendations will not guarantee that representative samples will be secured.

8. **Oil Analysis.** The condition of oil or grease affects machine reliability significantly. The chemical and physical properties of a lubricant have a direct effect on the lubrication situation. By contrast, the lubricant provides secondary information about the condition of the machine. Just as a blood test can reveal several illnesses in people, a thorough oil analysis can inform workers about several malfunctions within a machine. By using oil analysis on a regular basis, a testing baseline can be established for each piece of equipment. The oil analysis data can be trended, and deviation from the established baseline can be identified as an indication to take appropriate maintenance measures.

9. **Contamination Control.** At every point in time, there is danger that a lubricant will become contaminated by dust, dirt, water, or moisture unless your machines are working in a very clean environment. The first step toward contamination control is to set a target cleanliness level that takes into account the specific needs of the system. The current international standard for cleanliness of a lubricating fluid is defined by International Standards Organization (ISO) Standard 4406.

10. **Lubrication Training.** Companies that do not have a lubrication program at all believe that someone in the plant will lubricate the equipment when it needs it. They just need some grease guns and some oil drums around the site, and

people will add some oil when needed. Everyone involved in lubrication should have sufficient knowledge about his or her lubrication responsibilities from lubricator/technician to maintenance manager. To be effective, an education program must be delivered in conjunction with other resources that will allow individuals to absorb this new knowledge and introduce new behaviors into their daily work.

CONCLUSION

This book provides a comprehensive resource on the fundamental principles of lubricant selection and application, as well as an examination of which lubricants are most effective for specific applications. It also offers a detailed and highly practical discussion of lubrication delivery systems. You'll gain a clearer understanding of the "why" of relevant industrial lubrication practices and, importantly, how these practices facilitate optimized results. Also provided are expert tips on lubricant handling techniques, procedural setups, how and when to perform oil analyses, critical maintenance practices, equipment reliability issues, and more. The book combines lubrication theory with practical knowledge and provides many useful illustrations to highlight key industrial lubricant applications and concepts.

International Conference on Machine Learning (ICML) Questions

1. Run to failure is which type of maintenance strategy?
   a. Reactive
   b. Preventive
   c. Predictive
   d. Proactive

2. Maintenance done on calendar time is termed a _____ process?
   a. reactive
   b. preventive
   c. predictive
   d. proactive

3. Oil analysis belongs to what type of maintenance strategy?
   a. Reactive
   b. Preventive
   c. Predictive
   d. Proactive

4. Continuous improvement to maintenance activities can be achieved with which maintenance strategy?
   a. Reactive
   b. Preventive
   c. Predictive
   d. Proactive

5. The causes of failures and the consequence of such failures are important in which maintenance strategy?
   a. Reactive
   b. Preventive
   c. RCM
   d. Proactive

6. Which type of maintenance is most cost-effective?
   a. Reactive
   b. Preventive
   c. Predictive
   d. All of the above

CHAPTER 2

Friction, Wear, and Lubrication

The surfaces of machinery components appear well finished to the naked eye. However, when these surfaces are magnified, imperfections become apparent as hills and valleys called asperities. When dry surfaces move relative to one another, these asperities may rub, lock together, and break apart. The resistance generated by the rubbing of surfaces is called friction.

The energy expended in overcoming friction is dispersed as heat and is considered to be wasteful. This waste heat is a major cause of wear that leads to premature failure of equipment. Normally, the asperities of contact surfaces sometimes interlock to impede the sliding movement of machine parts. When the parts move, some of these asperities are deformed and may be subjected to very high localized temperatures. Given these high temperatures, the asperities cold flow or weld together and increase the resistance to motion. As the relative motion of surfaces continues, the welded asperities begin to shear off. In this process, small amounts of material (usually metal) are transferred from one surface to the other, and some small amount of material may be eroded from both metal surfaces. This gradual deformation and removal of material from solid surfaces is called wear.

Wear of metals occurs by plastic displacement of the surface material and by detachment of particles that form wear debris. The wear rate is a function of type of loading (e.g., impact, static, or dynamic), type of motion (e.g., sliding or rolling), and temperature. The primary function of a lubricant is to form a protective film between adjacent surfaces to reduce wear and to dissipate heat generated at these wear surfaces due to friction. The practice of lubrication is an ancient one. Water was probably the first lubricant. When primeval humans used water or ice to ease the sliding movement of heavy objects, the idea of lubrication was born. Fundamentally, lubrication is the reduction of friction to a minimum, replacing solid friction with fluid friction. In this chapter, we will discuss friction, wear, and lubrication and their impact on tribo systems.

FRICTION

Figure 2.1 shows a block sitting on a table and the frictional force acting on that block. To make the block slide on the table, a force F is needed to be applied to the block. Force F depends on the value of the load N, which is acting normal to the block.

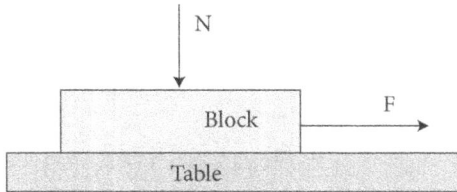

FIGURE 2.1   Example of frictional force acting on two surfaces.

The relation between F and N can be summarized by the equation

$$F = \mu N$$

where F is the frictional force, μ is the coefficient of friction, and N is the force acting perpendicular to the surface of contact.

The friction force depends on two factors:

1. The materials that are in contact and the nature of their surfaces. Rougher surfaces have higher coefficients of friction.
2. Force acting perpendicular to the surface of contact.

Two main types of friction are static and kinetic friction. The main difference between static and kinetic frictions is that static friction acts when the surfaces are at rest, and kinetic friction acts when there is relative motion between the surfaces. Kinetic friction can be classified into two types, solid and fluid friction. Solid friction acts in two different forms, sliding and rolling.

Sliding Friction

When two surfaces slide over each other without lubrication, sliding friction occurs. Here the two metal surfaces slide over each other in a dry state, and this causes high heat generation and sometimes wear of the surfaces. In due course, there will be complete failure of the machine parts. This type of friction takes place in a plain bearing or between a piston and a cylinder (Figure 2.2).

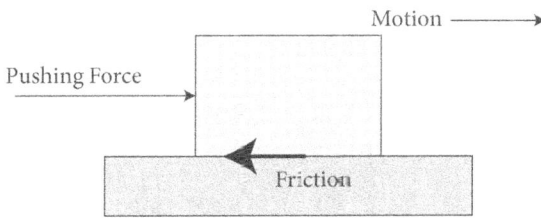

FIGURE 2.2 Sliding friction.

Laws of Sliding Friction

For Dry or Unlubricated Surfaces. Three laws govern the relationship between the frictional force F and the load or weight W of the sliding object for unlubricated or dry surfaces:

1. At lower pressures (normal force per unit area) the friction force is directly proportional to the normal load between the two surfaces. As the pressure increases, the friction force does not rise proportionally; but when the pressure increases significantly, the friction increases at a rapid rate until seizing takes place.
2. The coefficient of friction is independent of the area of contact, so long as the normal force remains the same. This is true for moderate pressures only. For high pressures, this law is modified in the same way as the first case.
3. At lower velocity, the friction force is independent of the velocity of rubbing. But as the velocities increase, the friction decreases.

For Lubricated Surfaces. The friction laws for well-lubricated surfaces are considerably different from those for dry surfaces, as follows:

1. If the surfaces are flooded with oil, the frictional resistance shall be independent of the pressure (normal force per unit area).
2. At low pressures the friction varies directly with the speed; but at high pressures, the friction is very great at low velocities.
3. For well-lubricated surfaces, the frictional resistance depends, to a great extent, on the temperature, because of two reasons. First because of change in viscosity of the oil and second because of journal bearings clearance with the shaft (the diameter of the journal bearing increases more rapidly than the diameter of the shaft with the rise in temperature).
4. If the bearing surfaces are flooded with oil, the coefficient of friction is independent of the nature of the materials that are in contact. As the lubrication

becomes less ample, the coefficient of friction becomes more dependent upon the type of material and its surface properties.

Rolling Friction

Rolling friction occurs when a cylindrical or spherical body rolls over another surface without lubrication, as in modern ball and roller bearings (Figure 2.3). There is less force to overcome in rolling friction than in sliding friction. However, when no lubrication is present, we can expect the same wear, heat, and eventual seizure of the contact surfaces in both instances but to a lesser degree in the case of rolling friction.

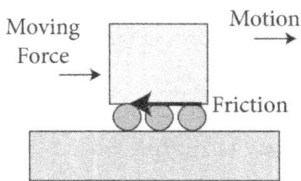

FIGURE 2.3   Rolling friction.

Ball and roller bearings are quite low in rolling friction because they use a very elastic and stiff material rolling in a very smooth cases made of a similarly stiff and elastic material. When loaded, the bearing surfaces do not deform very much, and because they have a high coefficient of elasticity, they return most of the energy to the surface rather than absorbing it and heating up, although they do get hot over time.

A plain bearing or bush bearing does not have rolling friction because the surfaces are sliding against each other, not rolling, even though one of the surfaces is rotating (Figure 2.4).

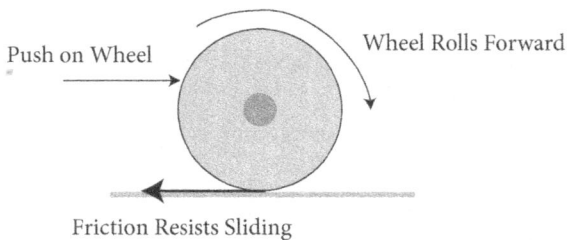

FIGURE 2.4   Static sliding starts the wheel rolling.

Properties of Rolling Friction

1. Unlike the coefficient of sliding friction, the coefficient of rolling friction varies with conditions and has a dimension expressed in units of length.
2. Ideally, a rolling sphere makes contact with a flat surface at a single point, and a cylinder will make contact along a line. In reality, though, the area of contact is slightly larger than a point or line because of elastic deformation of the surfaces. The friction is mostly attributed to elastic hysteresis. A perfectly elastic body may spring back immediately after relaxation of the deformation, but in reality a small amount of time is required to restore the object to its original shape. For this reason, the energy is not fully returned to the object but is retained and converted to heat. This energy acts as the rolling frictional force.
3. A small amount of slippage (sliding friction) occurs in rolling friction. Neglecting slippage, rolling friction is very small compared with sliding friction.

Laws of Rolling Friction

The laws of sliding friction cannot be applied to rolling bodies, but the following laws do apply:

1. The rolling friction force F is proportional to the load W and inversely proportional to the radius of curvature r, or $F = \mu W/r$, where μ is the coefficient of rolling resistance, in meters (inches). As the radius r increases, the frictional force decreases.
2. The rolling friction force F can be expressed as a fractional power of the load W times a constant k, where the constant k must be determined experimentally.
3. As the surface of rolling element becomes smoother, the friction force decreases.

Fluid Friction

Fluid friction is produced by fluids in motion or by contact between moving fluids and solids. Fluid friction is the force that resists motion either within the fluid itself or with another medium in contact with the fluid. The interactions between molecules of the fluid is called internal friction, and the interaction between the fluid and other matter is called external friction (Figure 2.5).

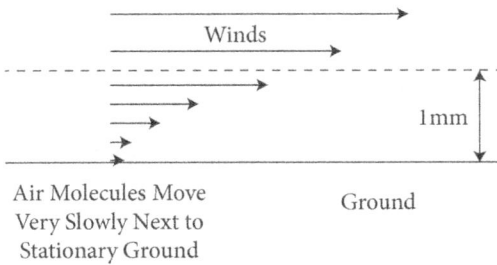

FIGURE 2.5   Fluid friction (in this case, air is the fluid).

The force needed to overcome fluid friction and move the top plate in Figure 2.6 at a velocity u is

$$F = \mu A(u/y)$$

where μ is fluid viscosity, A is the area of the top plate, u is the velocity of the top plate v_0, and y is the separation of the two plates d.

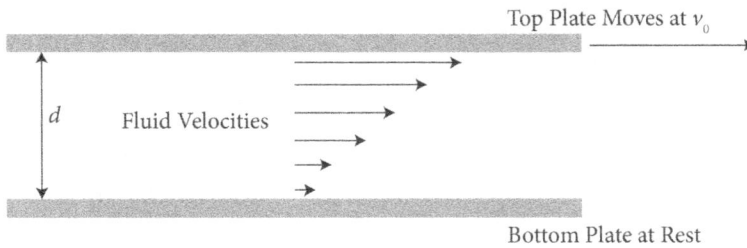

FIGURE 2.6   Overcoming fluid friction.

If we introduce a film of oil under pressure between the two surfaces, the hills and valleys are filled up by the particles of oil. When a sufficient number of these particles of oil are placed between the two surfaces, a thick strong film is produced. Then the hills and valleys slide by each other without interlocking. This is why some fluids are used as lubricants in between surfaces of sliding or rotating objects to reduce the friction coefficient. In this case, the contact surfaces become two sets of separate tribological surfaces, one with solid to fluid contact and the other with fluid to solid contact. Because the fluid friction is less in each set, the overall sliding friction decreases (Figure 2.7).

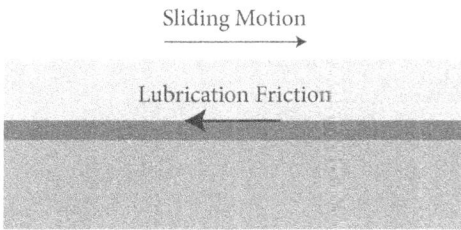

FIGURE 2.7   Layer of fluid lubricates the motion of the objects.

Ancient humans found that it required considerable effort to push or drag a heavy stone along the ground (Figure 2.8) and that it was much easier to roll the stone (Figure 2.9). Humans later discovered that a log was almost impossible to slide, as in Figure 2.10, but was easier to move when rolled, as in Figure 2.11. And when the log was placed in a river, as in

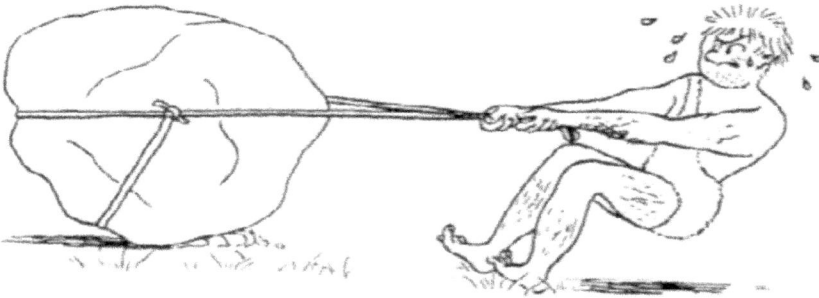

FIGURE 2.8   Sliding friction.

FIGURE 2.9   Rolling friction.

Figure 2.12, moving it was child's play by comparison. With the log in the river, we see fluid friction in action. The fact that the log is floating in the water is not the key point. When the water comes between the log and the bottom of the river the only force resisting the movement of the log is the resistance of one particle of water sliding over another. This is much less than the resistance encountered when the log was sliding in direct contact with the ground. Therefore, we actually reduce friction to a minimum by substituting fluid friction in place of solid friction.

FIGURE 2.10   Sliding friction—most effort.

FIGURE 2.11   Rolling friction—less effort.

FIGURE 2.12   Fluid friction—least effort.

Examples of early human struggles with friction in one form or another abound. Out of those early struggles, through long years of scientific study, we have come to the present-day concepts of friction, wear, and lubrication.

WEAR

Wear is the result of material removal by physical separation as a result of microfracture, chemical dissolution, or melting at the contact interface. Because of wear, all machines ultimately lose their durability and reliability. Therefore, wear control has become a necessity for the advanced and reliable technologies of the future.

For design of tribological systems and selection of materials based on their wear properties, an understanding of different wear modes, wear rates, and wear mechanisms is essential. The different modes of wear are adhesive, abrasive, fatigue, and corrosive. In general, wear does not take place via a single wear mechanism, so understanding the type of wear mechanisms in each mode of wear is important. With a small change in the tribological system, which is composed of dynamic parameters, environmental parameters, and material parameters, the wear rate can change dramatically. The dominant wear mode may change from one case to another as a result of changes in surface material properties, chemical film formation, and dynamic surface responses caused by frictional heating and wear.

Wear Mechanisms

Although wear processes are the result of mechanical loading, chemical processes also may be superimposed on them interfacially to influence such loading. The four wear mechanisms are described in the following subsections. Although each wear mechanism can occur alone, examination of wear damage indicates that there is nearly always a superposition of several wear mechanisms, which complicates any wear analysis and wear prevention. The four principal wear mechanisms acting alone and in combination under boundary lubrication conditions are described next.

Adhesion

Adhesion is a material interaction of tribological pairs. Because of shearing at points of contact of tribological pairs, asperities undergo adhesion or cold welding, resulting in adhesive wear. Every tribological surface possesses a certain degree of roughness, even though it is not visible to the naked eye. When two surfaces come into contact, only the roughness peaks of the two bodies are directly in tribological contact. As a result, the actual contact area is considerably smaller than the geometric area of the bodies in contact. Very high mechanical stresses occur in these microcontacts, but these stresses may be further increased by the relative movement of the bodies. Both elastic and elastoplastic deformation of the roughness peaks may then take place. Because of this deformation, the adhering adsorption and reaction layers are destroyed.

Normally, adhesion is likely a secondary mechanism in boundary lubrication because of the difficulty of achieving the proper lubrication conditions. Under effective lubrication conditions, adhesion of material may result by plowing between one bearing surface and the other surface. This may happen when a trapped third body, such as a fatigued asperity wear particle, is deformed plastically within the bearing surfaces.

As a countermeasure, chemically inert materials should be chosen for the tribological body surface. Material pairs with different types of atomic bonds have also proved to be favorable because interactions take place preferentially between materials having the same type of bonding.

Corrosive Wear

Corrosive wear occurs as a result of a chemical reaction on a wearing surface. The reactivity of material surfaces and the interaction with the outer environment may lead to a chemical reaction that can generate undesirable phenomena such as oxidation and corrosion. Corrosion products, usually oxides, have shear strengths that are different from those of the wearing metal surface from which they were formed. The oxides formed have a tendency to flake away and cause pitting of the surfaces. Ball and roller bearings require extremely smooth surfaces to function properly with minimum friction. Hence, corrosive pitting, if it occurs on the surface, is detrimental to these bearings. Sometimes corrosion boundary films such as soaps or fatty acids and other salts are formed by reaction with a lubricant additive or with lubricant oxidation products. If the films are pushed aside rather than entering the real load-supporting zones between asperities or dissolve in the bulk lubricant, then corrosive wear occurs.

In order to prevent tribological chemical reactions and corrosion from occurring, materials should be chosen that are chemically inert relative to the ambient medium and intermediate material. A suitable additive in the lubricant can prevent tribological chemical reactions from occurring via the formation of a protective layer on the surface of the material.

Abrasive Wear

Abrasion occurs predominantly in systems in which the tribological elements possess widely different hardness values. When two materials with significant hardness differences rub against each other, because of the penetration of the asperities of the hard material into the surface of the weaker material, plastic deformation takes place. As a result of the relative motion, sometimes due to stronger attack, the material breaks away from the surface, causing an extensive amount of wear. The process of material abrasion consists of such different processes as microplowing, microcutting, microfracture, and microfatigue that usually occur simultaneously.

Plowing. When material is displaced to the side, plowing takes place. In this process, material is taken away from the wear particles, resulting in the formation of grooves that do not involve direct material removal. The displaced material forms ridges adjacent to the grooves, and these ridges may be removed by subsequent rubbing with abrasive particles.

Cutting. This is the process that takes place when material is separated from the surface in the form of primary debris or microchips. In this process, no material is displaced to the sides of the grooves. This mechanism closely resembles conventional machining.

Fracture. This occurs when material is separated from a surface by a cutting process and the indenting abrasive causes localized fracture of the wear material. These cracks then freely propagate locally around the wear groove, resulting in additional material removal by spalling.

Fatigue. This is a process in which, because of cyclic loading, the surface material is weakened. When the wear particles are detached by the cyclic growth of microcracks on the surface, fatigue wear occurs.

Surface Fatigue Wear

Contact surfaces act under different loads in most machine elements such as gears, bearings, and friction drives. These elements are subject to repeated cycles of loading and unloading that lead to the initiation of cracks in their contact area and failure. This type of failure is termed fatigue failure because it is related to stress contact. Fatigue wear involves fracture of asperities or film material from repeated high stress. Commonly, this wear mechanism is evidenced by micropitting in rolling-element bearings. In sliding bearings, observation of fatigue wear of asperities is usually obscured by abrasion, denting, and adhesion produced by the trapping of spalled particles between the surfaces.

Wear fatigue is generated via the transfer of stress to the contact spots and the development of deformation in cycles. Then, after crack nucleation at the subsurface, cracks are generated and propagated until the creation of wear particles occurs. For example, when a ball bearing under a load rolls along its race, a different type of deformation occurs. The bearing is flattened somewhat, and the edges of contact are extended outward. This repeated flexing eventually results in microscopic flakes being removed from the bearing. But gear teeth frequently fail because of pitting, which occurs as a result of fatigue wear from the sliding motion.

Protection against surface fatigue is offered by mechanical surface treatment processes such as shot blasting or thermochemical diffusion processes such as nitriding or carbonization. Polished surfaces show better stability against surface fatigue because they exhibit fewer notches that would represent potential regions for the formation of cracks.

How to Avoid Wear of Lubricated Contacts

Application of lubricants changes the wear problem in contacting surfaces. Chemical wear can be avoided if the composition of the lubricant is properly selected. Abrasive wear will not take place as long as there is no accumulation of wear particles in the contacting interface. The internal features of materials such as their mechanical properties and chemical reactivity must be considered during the design stage to provide resistance to wear. External factors such as working environment, direct contact, the presence of oxygen, and temperature variations are also important. The most common property that affects the wear behavior of a material is hardness, and generally, the coefficient of wear is determined experimentally by considering the effect of temperature and load.

LUBRICATION

The best way to reduce friction and wear is to prevent asperity contact. Reduced friction and wear are achieved by inserting a lower-viscosity (shear strength) material between wearing surfaces. A material that is used to reduce friction and wear is called a lubricant. Lubricants are available in liquid, solid, and gaseous forms and flow between contact surfaces to form a film. Under the best of conditions, the moving parts do not actually make contact but rather glide on this film. Friction is greatly reduced because the resistance to movement is determined primarily by the viscosity of the lubricant. As friction is reduced, the amount of heat generated decreases greatly. Because heat and wear are associated with friction, their effects can be minimized to an acceptable level by reducing the coefficient of friction between the contacting surfaces.

Theory of Oil Film Formation

Now let's see what happens when we actually introduce a film of oil into a plain journal bearing. Figure 2.13 shows a journal bearing with a fluid film at rest. When the journal has been at rest for a while, the oil film between it and the bearing is squeezed out at the bottom, and the load is carried on the metal surfaces at the point of contact A in Figure 2.14a.

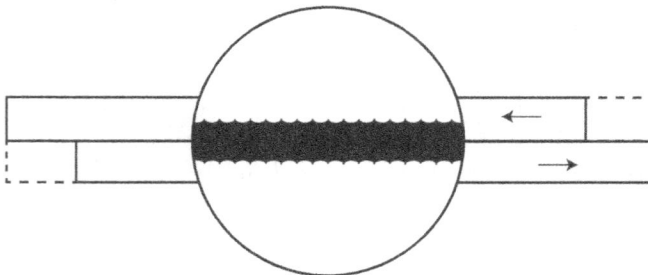

FIGURE 2.13   Magnified bearing surfaces with a fluid film.

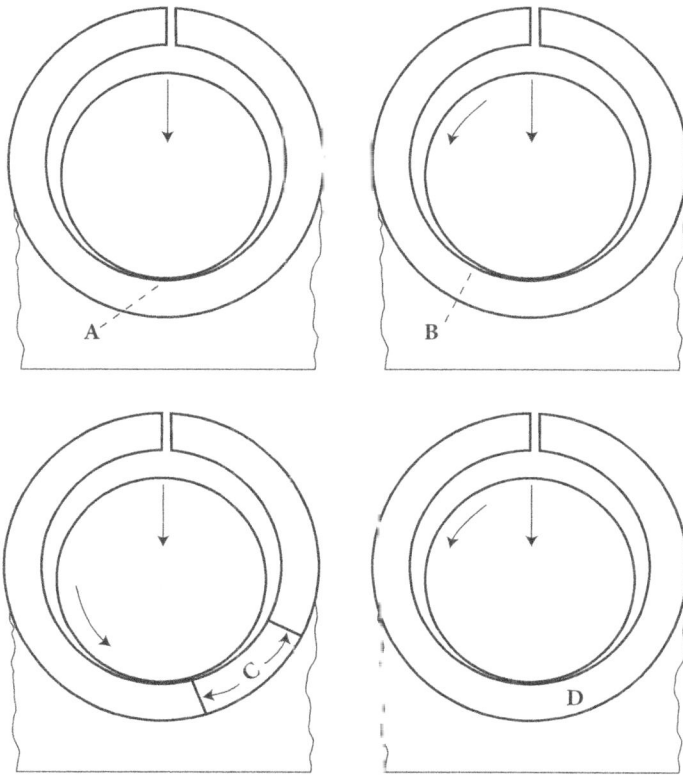

FIGURE 2.14   Oil film formation in a journal bearing.

As the journal starts to rotate (Figure 2.14b), it climbs up the bearing surface in a direction opposite to rotation. The layer of oil next to the slowly revolving journal clings to the journal and rotates with it. This layer of oil is dragged into the converging space between the journal and the bearing and begins to form a tiny film of oil at point B. The journal rotates with a thin film of oil until sufficient oil has been dragged into the converging space between journal and bearing to separate the surfaces (Figure 2.14c).

Notice that here we have an action similar to the action of the layers of oil between the flat bearing surfaces shown in Figure 2.15. The oil next to the journal clings to the journal and rotates when the journal rotates. The oil next to the bearing clings to the bearing and remains stationary. The layers of oil in the center slide between these two outside layers, with the oil closest to the journal moving the most and the oil closest to the bearing moving the least.

As speed increases, the wedging action of the lubricant moves in the direction of rotation and becomes stronger, lifting the journal into the position shown in Figure 2.14c. Then, as can be seen in the figure, the journal is riding on a film of oil, and the lubrication is

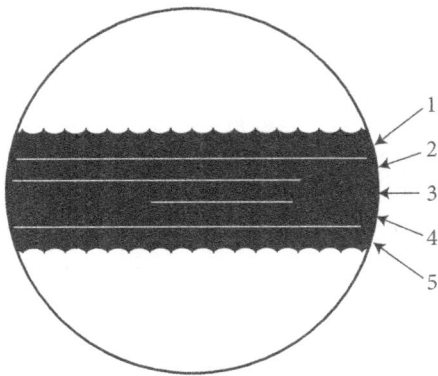

FIGURE 2.15   Oil film and wedge formation.

functioning perfectly. When the bearing is riding on a full film of oil, the hills and valleys on the two metal surfaces are completely separated, and all the solid friction is gone, replaced by fluid friction.

In describing this buildup of a film of oil in a plain journal bearing, we have illustrated the oil-wedge theory of lubrication. The term oil wedge is used because of the shape of the oil film. The oil next to the revolving journal forms the point of the oil wedge and pulls the adjacent layers of oil into the wedge-shaped clearance between journal and bearing. This oil wedge theory is widely accepted today as the basic principle of film formation in plain journal bearings. You can see in Figure 2.14 that the journal is not running in the center and that the oil film that surrounds the journal is not uniform in thickness. The point where the film is thin is known as the high-pressure area and is shown as C in Figure 2.14d. On top, where the film is thickest, we have the low-pressure area. Oil that is introduced into this low-pressure area is carried around with the shaft, is forced into a thin film at the high-pressure area, and has a lifting effect to support the journal in the position you see in the figure. Oil should always be introduced into a bearing in the low-pressure area.

What is an ideal oil film, and how does it govern good or poor lubrication in a bearing? An ideal oil film is a film of oil with sufficient body to keep the two metal surfaces separated under the speeds and loads imposed on the bearing. However, the oil must not be so heavy bodied that the internal friction from the oil itself causes excessive heating and wasted power. We will discuss different lubrication regimes in next section.

Lubrication Regimes

Lubricants are usually divided into four basic classes: oils, greases, solid lubricants, and gases. The most commonly used lubricants are mineral oils or synthetic fluids, but sometimes other types are also used, such as polytetrafluoroethylene (PTFE) as a solid lubricant for use in dry bearings, greases for use in rolling-element bearings, and gases such as air for use in gas bearings. To provide the machine elements with satisfactory life, the physical and chemical interactions between the lubricant and the lubricating surfaces must be fully understood. Before discussing the features that distinguish the four lubrication regimes from one another, let us discuss the geometry of the contacting bodies and how that affects lubrication.

Conformal and Nonconformal Surfaces
Depending on the gross geometry of the contacting bodies, lubrication processes can take many different forms. Other parameters such as roughness and texture of the sliding surfaces, the contacting load, pressure and temperature, the rolling and sliding speeds, the environmental conditions, the physical and chemical properties of the lubricant, and the material composition also play important roles. The two basic geometries for lubricated surfaces are conformal and nonconformal.

Figures 2.16 and 2.17 show the two distinctly different geometries. Figure 2.16 is a typical journal bearing; Figure 2.17 is a rolling-element bearing. Conformal surfaces fit together with a high degree of geometric conformity so that the load is carried over a relatively large area. The signifying characteristics of journal bearings—as well as other sliding-surface bearings such as thrust-pad bearings—are a high degree of conformity between the surfaces, relatively large effective contact areas, and low unit loading. By contrast, rolling-element bearings have poor conformity between surfaces, very small contact areas, and very high unit loads.

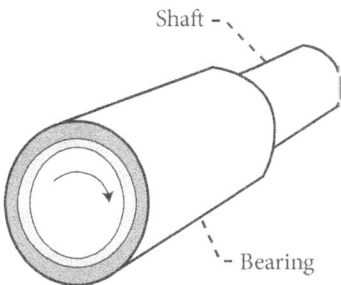

FIGURE 2.16   Conformal surfaces. (From Hamrock and Anderson 1983.)

Conformal surfaces usually appear in fluid-film journal bearings, trust bearings, and seals. In conformal bearings, usually surfaces are separated by a thick oil film generated hydrodynamically by the surface velocities or hydrostatically by an externally pressurized lubricant. These surfaces usually operate in the regime of thick-film hydrodynamic or hydrostatic lubrication. The load-carrying surface area normally remains constant when the load is increased. The important lubricant properties for conformal surfaces are viscosity and the temperature-viscosity coefficient.

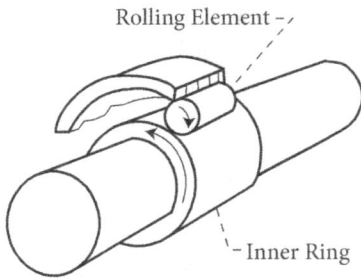

FIGURE 2.17   Nonconformal surfaces. (From B. J. Hamrock and W. J. Anderson.)

Nonconformal surfaces (Figure 2.17) have small lubrication areas because they do not conform geometrically to each other well. The full burden of the load is carried by a small lubrication area known as the Hertzian conjunction. This lubrication area enlarges with increasing load but is still small compared with the lubrication area of conformal surfaces. In these contacts, film thickness is very thin, and lubricant pressure is very high. Lubricant performance is affected strongly by the elastic deformation of the mating surface. The lubricant film thickness and pressure and the distribution of the film in the conjunction can be determined by elastohydrodynamic theories. Because of the extremely high pressure in the conjunction, the pressure-viscosity coefficient has a major effect in generating the lubricant film. Some examples of nonconformal surfaces are mating gear teeth, cams and followers, and rolling-element bearings.

Modes of Lubrication

Four different forms of lubrication can be identified for self-pressure-generating lubricated contacts:

- Hydrodynamic
- Mixed film
- Boundary
- Elastohydrodynamic (EHD)

Hydrodynamic Lubrication

In heavily loaded bearings such as thrust bearings and horizontal journal bearings, the fluid's viscosity alone is not sufficient to maintain a film between the moving surfaces. In these types of bearings, higher fluid pressures are required at the machine start to support the load until the fluid film is established. If this pressure is supplied by an outside source, then it is called hydrostatic lubrication. If the pressure is generated within the bearing by dynamic action, it is called hydrodynamic lubrication. Hydrodynamic lubrication is the more common type, and it is applicable to nearly all types of continuous sliding action where extreme pressures are not involved. Whether the sliding occurs on flat surfaces, as in most thrust bearings, or on cylindrical surfaces, as in journal bearings, the principle is essentially the same. Here the load-carrying surfaces are separated by a relatively thick film of lubricant. In this type of lubrication, metal-to-metal contact does not occur during steady-state operation of the bearing, and this is a stable regime of lubrication. Hydrodynamic lubrication is generally characterized by conformal surfaces. The lubricant pressure is self-generated by the moving surfaces drawing the lubricant into the wedge formed by the contacting surfaces. The wedge is formed by the high velocity of lubricant and generates the pressure to completely separate the surfaces and support the applied load.

In hydrodynamic lubrication, mechanical friction between moving surfaces is substituted by fluid friction. Hence there is no wear and no contact between the surfaces, and this mode is often referred to as stable lubrication. The distance between the two surfaces should be greater than the largest surface defect. The surface geometry is very important. The mating surfaces have to be such that a converging wedge of fluid can develop between the surfaces, which will allow hydrodynamic pressure of the lubricant to support the load of the shaft or moving surface. The film thickness is a function of three variables: viscosity, velocity, and load.

The distance between the two surfaces decreases with higher loads on the bearing, with less viscous fluids, and with lower speeds. As viscosity or velocity increases, the film thickness also increases. As the load increases, however, film thickness decreases. Viscosity, velocity, and operating temperature are also interrelated in hydrodynamic lubrication. If the oil viscosity is increased, the operating temperature will increase, and this, in turn, may try to reduce viscosity. If velocity increases, then temperature increases, which subsequently results in viscosity reduction.

The only friction present in a hydrodynamic lubrication system is the friction of the lubricant itself. In order to minimize friction, it would make sense to have a less viscous fluid, but too low a viscosity might jeopardize the system. Because viscosity is temperature dependent, special attention must be paid to preventing heating of the lubricant by the frictional force. Sometimes this is accomplished by passing the lubricant through a cooling reservoir to maintain the desired viscosity. Another way of handling the heat dissipation is to use viscosity index improvers, that is, additives to decrease the viscosity's temperature dependence.

Elastohydrodynamic Lubrication

Elastohydrodynamic (EHD) lubrication is characterized by nonconformal surfaces, where there is no asperity contact between the solid surfaces. It is a form of fluid-film lubrication where elastic deformation of the lubricated surfaces becomes significant. It is usually associated with highly stressed machine components such as rolling-element bearings and sliding surfaces under very heavy loads such as gears.

The formation of EHD lubrication in rolling-element bearings takes place as the oil wedge, similar to that which occurs in hydrodynamic lubrication, develops at the lower leading edge of the bearing. As the contacts approach, lubricant is forced from between contact surfaces because of the hydrodynamic effect. This flow is resisted by viscous forces, and there is an accompanying pressure rise that, in turn, raises the viscosity of the trapped lubricant. As the pressure increases, the surfaces deform elastically. The increased pressure resulting from the contact interaction represents load support by the contacts through the fluid film. The surface returns to its original configuration when the load is released.

In roller and ball bearings, the contact area is extremely small. Hence the force per unit area is extremely high. In roller bearings, load pressures may reach as high as 34,450 kPa (5,000 lb/in^2), and in ball bearings, load pressures may reach 689,000 kPa (1 million lb/in^2). It may seem that under these pressures, the oil is entirely squeezed away from the wearing surfaces, but under extremely high pressures, viscosity increases and prevents the oil from being entirely squeezed out. In the contact region at high pressures, the oil behaves virtually like a solid separating layer.

The EHD lubrication film thickness is a function of viscosity, surface roughness, and relative velocity between load-bearing surfaces. Higher viscosity and smoother surfaces increase film thickness. Engineers use the ratio of film thickness to surface roughness to estimate the life expectancy of a bearing system. In general, life expectancy is extended as the ratio increases.

Partial/Mixed Lubrication

In fluid-film lubrication as loads increase, lubricant is squeezed from between the parts, and the film becomes discontinuous. This regime is one of mixed-film lubrication. In the mixed-film lubrication regime, the behavior of the conjunction is governed by a combination of boundary and hydrodynamic or elastohydrodynamic effects. Some contact will take place between the asperities, and partial lubrication (sometimes referred to as mixed lubrication) will occur.

Between conformal surfaces, where hydrodynamic lubrication occurs if the film gets too thin, the mode of lubrication goes directly from hydrodynamic to partial. For nonconformal surfaces, where EHD lubrication occurs, if the film gets too thin, the mode of lubrication goes from EHD to partial. It is important to recognize that the transition from EHD to partial lubrication does not take place instantaneously as the severity of loading increases. Load carried by the pressures of the fluid that fills the space between the opposing solids slowly decreases. As the load increases, a larger part of the load is supported by the contact pressure between the asperities of the solids. If the pressures are too high or the running speeds are too low, the lubricant film will be penetrated and result in partial lubrication. In extreme environmental conditions, such as elevated ambient temperatures above 500°C (930°F) or vacuum environments, conventional liquid lubricants often become less effective because of their tendency to rapidly oxidize or decompose at elevated temperatures and their tendency to vaporize or creep away from lubricated surfaces under high vacuum leading to partial lubrication. The use of solid lubricants is often called for in these situations.

In the regime of mixed lubrication, because the load is shared between the lubricant pressure and the asperities, surface roughness and its orientation with respect to surface motion can have a significant influence on lubrication performance. In nonconformal contacts, roughness effects are very important because most of these contacts operate in mixed lubrication.

Boundary Lubrication

When a complete fluid film is not developed between the tribological surfaces, there may be momentary dry contact between wear surface asperities. Lubrication under these conditions is termed boundary lubrication. Boundary lubrication occurs when the oil supply is discontinuous or the loads on the surfaces are high enough to squeeze out the lubricant film and allow metal-to-metal contact. With heavy loads and low running speeds, where fluid-film lubrication is difficult to attain, boundary lubrication occurs. Lubricant will be present, but a continuous film is not present on the moving surfaces. The coefficient of friction is highest in this mode of lubrication. The friction and wear in boundary lubrication are determined by the interaction between the solids and between the solids and the liquid. The bulk flow properties of the liquid play little part in the friction and wear behavior. The most common example of boundary lubrication involves gear tooth contacts and reciprocating equipment. In addition, bearings that operate on fluid-film lubrication can experience boundary lubricating conditions during routine starting and stopping of equipment.

Also, thick-film lubrication deteriorates rapidly under combinations of extreme pressure, extreme temperature, low rolling speeds, and high sliding speeds. For these conditions, the lubricant film thickness is extremely small compared with the surface roughness, and the contacts operate in the regime of boundary lubrication. Straight mineral oils do not offer much relief from friction and wear. Additives must be relied on to modify friction, reduce wear, and reduce oxidation of the lubricant.

Lubricants employed under boundary lubrication conditions must possess an added quality referred to as oiliness or lubricity to lower the coefficient of friction of the oil between the rubbing surfaces. Oiliness is an oil enhancement property provided by the use of chemical additives known as antiwear (AW) agents. The opposite sides of the oil film have different polarities, like a magnet. When an AW oil adheres to a metal surface, the sides of the oil film not in contact with the metal surface have identical polarities and tend to repel each other and form a plane of slippage. Most oils intended for use in heavier machine applications contain AW agents.

Figure 2.18 shows the wear rate in the various lubrication regimes, as determined by the operating load. In the hydrodynamic and EHD regimes, there is little or no wear because there is no asperity contact. In the boundary lubrication regime, the degree of asperity interaction and the wear rate increase as the load increases. The transition from boundary lubrication to an unlubricated condition is marked by a dramatic change in wear rate. As the relative load is increased in the unlubricated regime, the wear rate increases until scoring or seizure occurs, and the machine element can no longer operate successfully. Most machine elements cannot operate long with unlubricated surfaces. Together Figures 2.19 and 2.20 show that the friction and wear of unlubricated surfaces can be greatly decreased by providing boundary lubrication.

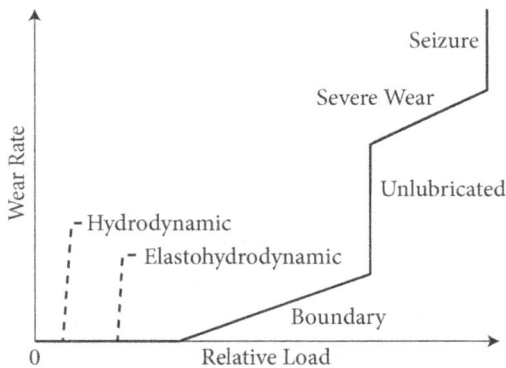

FIGURE 2.18   Wear rate of various lubrication regimes. (From B. Bhusan, *Introduction to Tribology*. New York: Wiley, 2002. Used with permission.)

Figure 2.19 shows a schematic of three general categories of fluid film.

Figure 2.20 illustrates the film conditions existing in fluid film and boundary lubrication. The surface slopes in this figure are greatly distorted for purposes of illustration. To scale, real surfaces would appear as gently rolling hills rather than sharp peaks. The surface asperities are not in contact for fluid film lubrication but are in contact for boundary lubrication.

Lubricant
Film

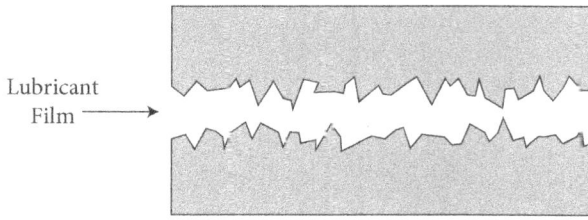

FLUID-FILM LUBRICATION
Surfaces well separated by
bulk lubricant film

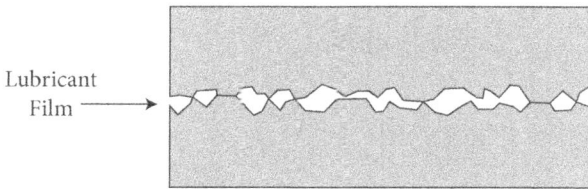

Lubricant
Film

BOUNDARY LUBRICATION
Performance essentially
depends upon the quality
of the boundary film

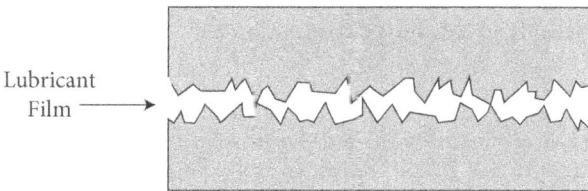

Lubricant
Film

MIXED-FILM LUBRICATION
Both the bulk lubricant and
the boundary film play a role

FIGURE 2.19   Three general categories of fluid film.

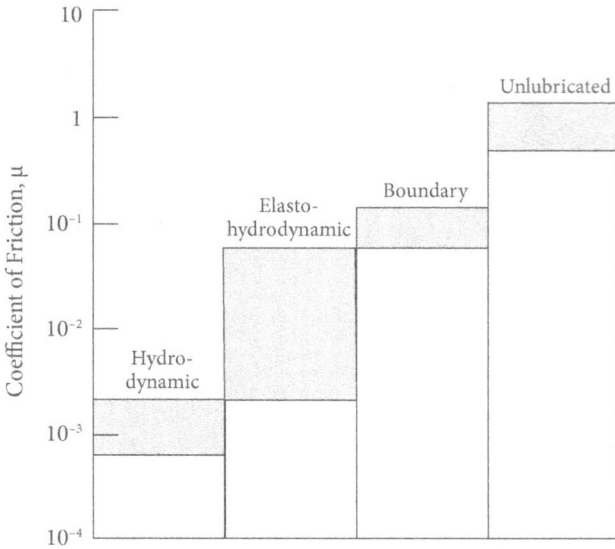

FIGURE 2.20   Bar diagram showing coefficients of friction for the four lubrication conditions.

In boundary lubrication, considerable asperity contact occurs, and the lubrication mechanism is governed by the physical and chemical properties of thin surface films, which are of molecular proportion (1–10 nm). The frictional characteristics are determined by the properties of the solids and the lubricant film at the common interfaces. Partial lubrication (sometimes referred to as mixed lubrication) is governed by a mixture of boundary and fluid-film effects. Most of the scientific unknowns lie in this lubrication regime.

Stribeck Curve

The physical and chemical interactions between the lubricant and the lubricating surfaces must be understood in order to provide the machine elements with satisfactory life. The Stribeck curve is used to understand the features that distinguish different lubrication regimes from one another. The pressure distribution in the lubricating film is composed of either a hydrostatic or a hydrodynamic (or both) contribution. Hydrostatic bearings do not require relative motion of the bearing surfaces to build up the pressure to support the load. They require an external pump to develop the pressure. But hydrodynamic pressure is generated owing to the sliding motion of the surfaces, which act as a pump to develop pressure. These lubrication regimes, which form lubricating films without an external pumping agency (self-acting), are found in the Stribeck curve.

A typical Stribeck curve is shown in Figure 2.21, which is often used to show zones or regimes of operation of a shaft in a bearing in terms of the effectiveness of lubrication.

Historically, the Stribeck curve was first widely disseminated because of Stribeck's systematic and definitive experiments that explained friction in journal bearings in the early 1900s (see Y. Wang, Q. Wang, C. Lin, and F. Shi, "Development of a set of Stribeck curves for conformal contacts of rough surfaces. Tribol. Trans. 49: 526–535, 2006; and X. B. Lu and M. M. Khonsari, "The Stribeck curve: Experimental results and theoretical prediction." J. Tribol. 128: 789–794, 2006).

Stribeck performed a series of experiments on journal bearings in which he measured the coefficients of friction of various surfaces as a function of load, speed, and temperature. Later, Hersey performed similar experiments and devised a plotting format based on coefficient of friction as a function of a dimensionless parameter ZN/p, where Z is the viscosity of the lubricant, N is the rotational speed of the shaft, and p is the nominal contact pressure between the shaft and bearing (Hersey number; see B. Bhusan, Introduction to Tribology. New York: Wiley, 2002). The Stribeck curve portrays the variations in friction over a range of the ratio ZN/p.

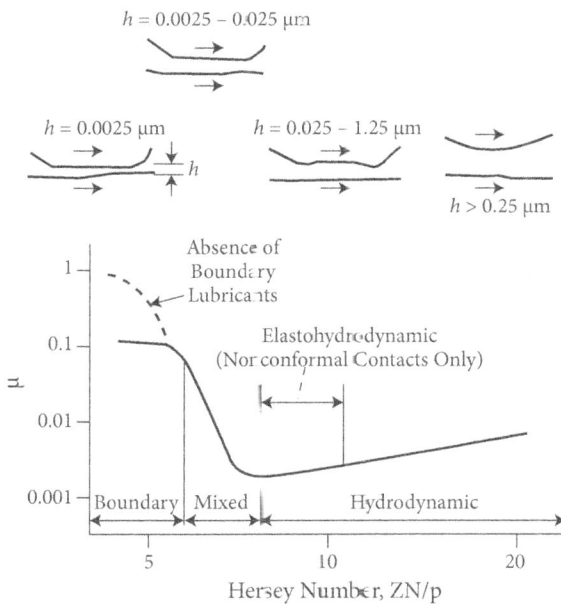

FIGURE 2.21 Stribeck curve showing the effect of Hersey number on coefficient of friction.

The ordinate in Figure 2.21 is the coefficient of friction under steady-state conditions. The abscissa in the figure is a dimensionless number, sometimes referred to as the Hersey number, and is given as ZN/p. A high Hersey number usually means a relatively thick lubricant film, whereas a small number means a very thin film. At an extremely low Hersey number, no real lubricant film can develop, and there is significant asperity contact, resulting in high friction. This regime is depicted in the left-hand section of the Stribeck curve. In this region, a high friction value with a low Hersey number represents the dominance of boundary lubrication region. Boundary lubrication occurs when the solid surfaces are so close together that surface interaction between solid asperities dominates the contact. In hydrodynamic lubrication as well, as the load increases, speed or fluid viscosity decreases, and the coefficient of friction can increase to high levels (~0.1 or higher), resulting in the boundary lubrication regime. The failure in boundary lubrication leads to adhesive and chemical wear.

As the Hersey number increases, there is a steep decrease in friction values. This occurs because of an increase in lubricant film thickness; as a result, the load is supported between the surface asperities and the pressurized liquid lubricant present in the conjunction. This region has widely varying friction values and strongly depends on operating conditions. This regime represents dominance of mixed-film lubrication. In this region, asperity collision is thought to occur because contacting bodies are expected to be supported on a combination of asperity-asperity contact points and fluid regions between asperities. In the mixed-film region, the friction is high and totally unpredictable. Mixed-film lubrication occurs between the transition of boundary lubrication and hydrodynamic lubrication regimes. There may be more frequent solid contacts, but at least a portion of the bearing surface remains supported by a partial hydrodynamic film.

As the Hersey number increases further, friction reaches a lower plateau value, indicating the onset of hydrodynamic lubrication. This normally occur at high speeds, high viscosities, and low loads, when the surfaces are completely separated by a thick lubricant film. Here the surfaces are effectively separated by the liquid lubricant, and asperity contact has negligible effect on load and friction. In the area of hydrodynamic lubrication, friction is determined by the rheology of the lubricant. Subsequently, the Stribeck curve shows a slight increase in friction with respect to Hersey number in the hydrodynamic regime. This occurs because as ZN/p increases, the viscous losses increase. Increased friction can be attributed to increased redundant work in the lubricant or the increases in shear strength.

A distinction is often made between EHD and full-film lubrication in a Stribeck curve. This is useful for some bearing types, gears, or cams, but it should be recognized that conformal contacts do not encounter EHD lubrication. Therefore, this regime may or may not be identified in the Stribeck curve depending on the source. For nonconformal concentrated contacts where loads are high enough to cause elastic deformation of the surfaces and high pressure in the conjuncture owing to viscosity effects, EHD lubrication occurs. Film thickness in this regime ranges from 0.025 to 1.250 μm. EHD lubrication is a subset of hydrodynamic lubrication in which the elastic deformation of the contacting solids plays a significant role in the hydrodynamic lubrication process. The film thickness in EHD lubrication is thinner than that in conventional hydrodynamic lubrication. In isolated areas, asperities may actually touch. Therefore, in these systems, boundary lubricants that provide boundary films on the surfaces for protection against any solid-solid contact are used. In EHD lubrication, in addition to adhesive and corrosive wear for heavily loaded contacts (rolling-element bearings and gears), fatigue wear is most common.

Factors That Affect the Effectiveness of the Oil Film

Factors that affect the oil film are viscosity, speed, and load, although the most important single factor that determines the effectiveness of an oil film is the viscosity of the oil:

Viscosity. This is a measure of an oil's resistance to flow. An ideal oil film on a bearing depends on selecting an oil with the right viscosity to allow the oil wedge action to raise the journal sufficiently off the bearing to maintain separation of the two metal surfaces.

Speed of the journal. The viscosity of the oil and the speed of the journal are important factors for maintaining a good oil film in a bearing. At slower journal speeds, higher-viscosity or thicker oil must be used. As journal speeds are increased, a thinner or lower-viscosity oil is needed. For the sake of clarity, low speed bearings usually have relatively wide clearances between the bearing and the journal. Therefore, heavy or high-viscosity oil is used. High-speed bearings usually have closer-fitting parts. Therefore, thin or low-viscosity oil is required.

Bearing load. Loading must also be considered because the oil must have sufficient viscosity or body to maintain a good oil film to support the load. We can see the effect of using the same oil under different loads: a 500-lb load squeezes the oil, but the viscosity of the oil is enough to keep the bearing surfaces apart. When a 1,000-lb. load is applied to the same oil, the squeezing action on the oil is greater. Therefore, it is reasonable to assume that the oil film will be thinner under the heavier load. If we want to maintain a proper oil film under a 1,000-lb load, we must use a heavier oil. The surfaces must be held far enough apart to keep those hills and valleys on the surfaces from interlocking. At the same time, the oil must not be so heavy that heat is produced from the internal friction of the oil itself.

CONCLUSION

While wear and heat cannot be completely eliminated between tribological pairs, they can be reduced to negligible or acceptable levels. Because heat and wear are associated with friction, both effects can be minimized by reducing the coefficient of friction between the contacting surfaces. This chapter deals with lubrication fundamentals and, in particular, defines the various lubrication mechanisms: hydrodynamic, elastohydrodynamic, mixed, boundary, and extreme pressure. An understanding of different lubrication regimes will help you to select the right lubricant and additives for different applications.

ICML Questions

1. The resistance generated by rubbing of surfaces is called
   a. friction.
   b. wear.
   c. heat.
   d. corrosion.

2. Rougher surfaces have higher coefficients of friction than smoother surfaces.
   a. True
   b. False
   c. Same coefficients of friction
   d. Cannot say

3. As long as the normal force remains the same, the coefficient of friction is
   a. independent of the area of contact.
   b. directly proportional to the area of contact.
   c. inversely proportional to the area of contact.
   d. two times.

4. At lower velocities, the friction force is independent of the velocity of rubbing. But as the velocity increases,
   a. the friction decreases.
   b. the friction increases.
   c. the friction remains the same.
   d. Cannot say.

5. For well-lubricated surfaces, the frictional resistance depends, to a great extent, on the temperature because
   a. of a change in the viscosity of the oil.
   b. of the journal bearing's clearance with the shaft.
   c. of friction force directly proportional to temperature.
   d. Both a and b

6. A plain journal bearing or bush bearing in which one of the surfaces is rotating
   a. has rolling friction.
   b. has sliding friction.
   c. has both rolling and sliding friction.
   d. has static friction.

7. In rolling friction, the friction force F
   a. is inversely proportional to the radius of curvature r,
   b. is directly proportional to the radius of curvature,
   c. has no effect on the radius of curvature,
   d. is within the same condition as if one object is under sliding friction, rolling friction and fluid friction,

and then

   e. sliding friction > rolling friction > fluid friction.
   f. sliding friction < rolling friction < fluid friction.
   g. rolling friction < fluid friction < sliding friction.
   h. sliding friction < rolling friction < fluid friction.

8. A typical journal bearing has a form of
   a. conformal surfaces.
   b. nonconformal surfaces.
   c. None of the above

9. Thrust bearings act on
   a. hydrostatic lubrication.
   b. hydrodynamic lubrication.
   c. mixed-film lubrication.
   d. None of the above

10. Hydrodynamic lubrication is generally characterized by
   a. conformal surfaces.
   b. nonconformal surfaces.
   c. both types of surfaces.
   d. any type of surface.

11. In hydrodynamic lubrication, friction between moving surfaces is
   a. mechanical friction.
   b. fluid friction.
   c. sliding friction.
   d. All of the above

12. In hydrodynamic lubrication, the film thickness is a function of
   a. viscosity.
   b. velocity.
   c. load.
   d. All of the above

13. In hydrodynamic lubrication, as viscosity increases,
   a. film thickness also increases.
   b. film thickness decreases.
   c. film thickness remains the same.
   d. film thickness may increase or decrease.

14. In hydrodynamic lubrication, as velocity increases,
   a. film thickness also increases.
   b. film thickness decreases.
   c. film thickness remains the same.
   d. Film thickness may increase or decrease.

15. In hydrodynamic lubrication, as the load increases,
   a. film thickness also increases.
   b. film thickness decreases.
   c. film thickness remains the same.
   d. film thickness may increase or decrease.

16. EHD lubrication is characterized by
   a. conformal surfaces.
   b. nonconformal surfaces.
   c. both types of surfaces.
   d. any surface.

17. In EHD lubrication, film thickness is a function of
   a. viscosity.
   b. surface roughness.
   c. relative velocity.
   d. All of the above

18. Boundary lubrication occurs when
   a. there are heavy loads.
   b. there are low running speeds.
   c. where fluid-film lubrication is difficult to attain.
   d. All of the above

19. ZN/p, where Z is the viscosity of the lubricant, N is the rotational speed of the shaft, and p is the nominal contact pressure between the shaft and bearing, is called
  a. the Stribeck number.
  b. the Hersey number.
  c. the Einstein number.
  d. the Newton number.

20. A low Hersey number represents the dominance of
  a. a mixed-film lubrication region.
  b. a boundary lubrication region.
  c. a hydrodynamic lubrication region.
  d. an EHD lubrication region.

21. As the Hersey number increases, there is
  a. a steep decrease in friction values.
  b. a steep increase in friction values.
  c. No change in friction values
  d. All of the above

CHAPTER 3

Lubricants and Their Properties

Lubricants are substances that are introduced between two rubbing surfaces to reduce friction and the heat generated when the surfaces move. The idea of using a lubricant to reduce friction and wear has been known since ancient times, perhaps when the wheel was discovered.

Although lubricants are used to reduce friction and wear, they also have some secondary functions, including:

- **Cooling.** If metals rub against each other, the friction force generates heat in the contact area that needs to be absorbed or released. Lubricants are applied to reduce friction, absorb heat from the contact surfaces, and transport it to locations where heat can be dispersed safely.
- **Sealing.** Sealing involves closing the gap between systems such as the space between pistons and cylinders in the internal combustion engine or air compressor blocks. Lubricants act as a dynamic seal in locations such as the piston ring and cylinder to prevent the leakage of combustion gas and the inflow of external foreign substances to maintain the defined internal pressure and protect the system.
- **Cleaning.** Long-term use of lubrication systems may lead to corrosion or aging, producing foreign substances such as sludge from deterioration. The lubricant itself cleans out foreign substances such as soap and prevents contaminants from adhering to components. The detergents and dispersants play a key role in maintaining internal cleanliness
- **Rust prevention.** Metals produce rust when they come into contact with water and oxygen. However, rust formation can be controlled and system lifetime can be extended if metal surfaces are coated with a lubricating film. The lubricant also prevents or minimizes internal corrosion by setting up a barrier between components and corrosive materials.

■ **Dampening shocks.** Some components, such as gears or bearings, have point or line contact between surfaces. If the load is increased instantaneously, components may suffer damage as a result of shock loading. A lubricant can cushion the blow of mechanical shock by forming an oil film to distribute the load on the film. As a result, the mechanical shock to components is dampened, and wear is minimized.

■ **Condition monitoring.** Analysis of used lubricating oil not only indicates the condition of the oil but also provides a fast and accurate picture of what is happening inside the machinery. Oil condition monitoring acts as a predictive maintenance tool to help clients avoid costly machine failures. In oil condition monitoring, machinery wear, lubricant quality degradation, and other problems can be diagnosed.

Lubricant at the interface of two parts reduces adhesive friction by lowering the shear strength of the interface. Lubricants are classified into four types based on their molecular state (see Figure 3.1):

1. Solid lubricants
2. Semisolid lubricants
3. Liquid lubricants
4. Gaseous lubricants

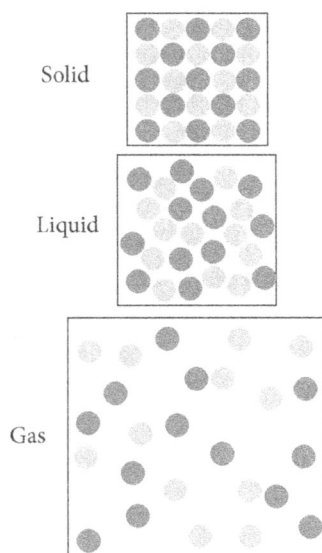

FIGURE 3.1 Molecular states of lubricants.

SOLID LUBRICANTS

Normally, solid lubricants that have a layered crystalline structure are interposed as a film between sliding and/or rolling surfaces to reduce friction. Some layer-lattice solids are graphite, molybdenum disulfide, boron nitride, and borax. Because of the layered crystalline structure, they shear more easily under a given load and are able to reduce friction between two surfaces sliding against each other. Certain solids with nonlamellar structures also function equally well as dry lubricants in some applications. These include certain soft metals (e.g., lead, silver, tin), polytetrafluoroethylene (PTFE), some solid oxides, rare earth fluorides, and even diamond. Solid lubricants are normally used in the form of powders or as bonded solid films.

For suitability of a material to be used as a solid lubricant, the following properties are important:

1. **Low shear strength in the sliding direction.** This is the property that provides a low coefficient of friction of the lubricant as a result of easy shear movement of the material.
2. **High compression strength perpendicular to the sliding direction.** The solid lubricant should possess high compression strength and should be capable of withstanding high loads without sufficient direct contact between the rubbing surfaces.
3. **Good adhesion of the solid lubricant to the substrate surface.** Solid lubricants must have properties that enable them to create a film and adhere to the machine surface even at high shear stresses.
4. **Thermal stability.** One of the most significant uses of solid lubricants is in high-temperature applications where other lubricants are unsuitable for use. Hence solid lubricants should have good thermal stability that resists undesirable phase or structural changes at temperature extremes.
5. **Oxidation stability.** Within applicable temperature ranges, a solid lubricant should not undergo undesirable oxidative changes.
6. **Chemical reactivity.** A solid lubricant should form a strong, adhesive film on the base material.
7. **Mobility.** Mobility of adsorbents on the surfaces promotes self-healing and prolongs the endurance of films. If the film remains intact, the life of solid films can be maintained longer.
8. **Melting point.** During service, a solid lubricant should not exceed its melting point; otherwise, the atomic bonds that maintain the molecular structure will be destroyed, rendering the lubricant ineffective.

9. **Hardness.** Solid lubricants should not have excessive hardness. A maximum hardness of 5 on the Mohs scale is a good limit for solid lubricants.

10. **Chemically stable.** At operating and environmental conditions, a solid lubricant should be chemically stable and prevent corrosion.

Classification of Solid Lubricants

The most commonly used solid lubricants are:

1. Inorganic compounds
2. Soft metals
3. Polymeric materials (PTFE)
4. Ceramics

Inorganic Lubricants with Lamellar Structure

These materials have a layered structure consisting of hexagonal rings forming thin parallel planes. The planes are bonded to each other by weak forces, and the atoms within the planes are strongly bonded. The layered structure helps in sliding movement of the parallel planes. The weak bonding between the planes helps in creating lubricating properties of the materials.

The most common inorganic solid lubricants are graphite, molybdenum disulfide (MoS_2), and boron nitride (BN).

Graphite.   Graphite is made up of planes of polycyclic carbon atoms that are hexagonal in orientation. Carbon atoms in the planes have a hexagonal orientation, and the long distance of carbon atoms between planes makes the bonding weaker. For graphite lubrication, water vapor is a necessary component. The adsorption of water by graphite reduces the bonding energy between the hexagonal planes of the graphite to a level that is less than the adhesion energy between a substrate and the graphite. Because water vapor is a requirement for lubrication, graphite is most suitable for lubrication in air and is not effective in vacuum. Graphite can promote galvanic corrosion because it is electrically conductive.

Graphite has high thermal and oxidation stability at temperatures up to 500–600°C, which enable its use at high temperatures and high sliding speeds. But high temperatures and sliding speeds may cause failure, and these can be neutralized by the use of suitable additives. Graphite is effective at temperatures up to as high as 450°C in an oxidative atmosphere. However, the wear rate increases with temperature in dry conditions.

Carbon graphite seals are the most common example of graphite used as a solid lubricant. These seals provide low friction but a tight seal and transfer layers of graphite on mating surfaces.

The primary limitations of graphite are its low tensile strength and lack of ductility. Mechanical distortion of graphite limits its use to moderate loads (<275 MPa). It is interesting to note that the presence of water helps graphite in lubrication, whereas the presence of water is detrimental to MoS_2. By contrast, vacuum is detrimental to graphite but favorable to MoS_2.

Uses. Graphite is used as lubricant in air compressors, the food industry, railway track joints, brass instrument valves, open gears, ball bearings, and machine-shop works. It is commonly used for lubricating locks because a liquid lubricant allows particles to get stuck in the lock, worsening the problem.

Molybdenum Disulfide (MoS₂). MoS_2 is one of the most commonly used solid lubricants. It has a hexagonal crystal structure that shears easily. The lubrication performance of MoS_2 is often better than that of graphite and is most suitable in vacuum applications (where graphite is not suitable).

MoS_2 starts to oxidize above 350°C in oxygen and 450°C in air, but the main oxidation product is molybdic oxide, which is itself a fair high-temperature lubricant. In high vacuum, the disulfide is said to be stable up to 1,000°C, and it evaporates very slowly, so it has been widely used in space. It has high load-carrying capability (>700 MPa), low friction, resistance to high temperature, particularly in space. Its weakness, however, is moisture, which degrades its performance.

Uses. MoS_2 acts as a lubricant at high temperatures and in oxidizing atmospheres, where liquid lubricants typically will not survive. A typical application includes fasteners that are easily tightened and unscrewed after a long stay at high temperatures. It is also used in constant-velocity (CV) joints and space vehicles. It does lubricate in a vacuum.

Boron Nitride (BN).   Hexagonal BN is a ceramic powder lubricant. It has high thermal conductivity and high temperature resistance up to 1,200°C in an oxidizing atmosphere.

Uses. Hexagonal BN is also called *white graphite*, and it is used in space vehicles.

Soft Metals

Soft metals with low hardness are used as solid lubricants in key sliding and rolling mechanical components for reducing friction and improving antiwear ability. Some soft metals possess lubrication properties because of their low shear strength and high plasticity, such as lead (Pb), tin (Sn), bismuth (Bi), indium (In), cadmium (Cd), and silver (Ag). These metals can provide effective isolation on the material surfaces, thereby reducing friction and providing a lubrication effect that prevents seizure.

The benefits of soft-metal lubricants include:

- High-temperature stability
- High load-carrying capacity

Soft metals are used in pure form or as alloys or in the form of coatings such as lead- and tin-based engine bearing overlays. Soft-metal lubricant coatings are produced by electroplating and thermal spraying. However, these lubricants are not suitable for use with stainless steel at temperatures above 1,000°C because of the potential for galvanic corrosion.

Uses. Soft metals are widely used as solid lubricants in engine bearings.

Polymeric Materials

Polytetrafluoroethylene (PTFE) is a typical example of an organic lubricant with a chain structure of polymeric molecules. The molecular structure of this material consists of long-chain molecules parallel to each other. The bonding strength between the molecules is weak, and thus they may slide past one another at low shear stresses. Because of strong bonding between the atoms within the molecules, their strength along the chains is high, which provides good lubrication properties. Poor adhesion of PTFE to other materials is responsible for very low coefficients of static and dynamic friction (down to 0.04). Operating temperatures are limited to about 260°C. PTFE has high chemical stability and great chemical inertness because of the carbon–fluorine bonds. Teflon, a commonly used PTFE, is the trade name of Du Pont.

The weaknesses of PTFE are that it is too soft and has a high wear rate, poor creep resistance, poor thermal conductivity, high thermal expansion, a temperature limit of 250°C, and vacuum is detrimental to its performance.

Uses. PTFE is widely used as an additive in lubricating oils and greases. It is used in very light load applications. Because it is nontoxic, it is useful in the pharmaceutical and food industries.

Ceramics and Cermets (Metal-Bonded Ceramic) Coatings

Metals are not capable of withstanding the incredibly high temperatures typically encountered in airplane jet engines and space rockets. Hence ceramics are used at high temperatures and are able to resist attack by chemicals and such things as oxygen in the air. But metals are fairly poor when it comes to doing interesting things like conducting electricity or heat or bending and flexing.

Sometimes particles of the metal are attached to a ceramic base for used in electrical applications. Electrical components can get extremely hot, and they need to behave like ceramics, but because they also need to conduct electricity, it helps if they work like metals. *Cermet*, a ceramic plus metal composite, is the generic name for a whole range of different composites.

Ceramics and cermets can be used in applications where a low wear rate is more important than low friction. These composites can be used at temperatures up to 1,000°C. Ceramic/cermet coatings up to 0.5 mm thick on metal substrates offer excellent wear resistance of metals at a minimum cost.

Uses. Ceramic wet chain lube is a unique ceramic coating that provides incredible durability and long-distance performance on all bicycle chains. Ultra-low-viscosity bearing lubricant is specially formulated to keep the ceramic bearings rolling faster and longer. Microceramic lubricant contains advanced ceramic nanoflakes, which are used for aerospace applications. Ceramic nanoflakes are also used for reducing machine noise. For components such as resistors and vacuum tubes (valves), cermets offer a perfect solution.

Advantages of Solid Lubricants

1. Solid lubricants are more effective at high loads and speeds than liquid lubricants.
2. They are highly stable in extreme temperatures, pressures, and reactive environments.
3. They do not require lubrication distribution systems and seals and thus allow equipment to be lighter and simpler.
4. They have high resistance to deterioration in storage.

Disadvantages of Solid Lubricants

1. **Poor self-healing properties.** The useful life of a solid lubricant tends to get shortened by a broken solid film.
2. **Poor heat dissipation.** In polymers, poor heat dissipation occurs because of their low thermal conductivity.
3. **Higher coefficient of friction and wear.** Compared with hydrodynamically lubricated bearings.
4. **Color.** The colors associated with solid lubricants may be undesirable.
5. **Resistance to systemization.** Solid lubricants are difficult, if not impossible, to feed into a lubrication system.
6. **Thermal expansion mismatch.** Solid lubricants can have very different coefficients of thermal expansion from the metals that are being lubricated. Therefore, desirable mechanical clearance will vary during operation at elevated or reduced temperatures.

SEMISOLID LUBRICANTS (GREASES)

Greases are defined as semisolid materials produced by the dispersion of a thickening agent in a liquid lubricant. The liquid does the lubricating, whereas the thickener primarily holds the oil in place and provides resistance to flow. One of the important functions of grease is to remain in contact with moving surfaces and not leak out under gravity or centrifugal action. It also should not be squeezed out under pressure. The major practical requirement of a grease is that it must retain its properties under shear at all temperatures to which it may be subjected during use. At the same time, the grease must flow into the bearing with ease via grease guns and from one point to another in the machine as needed. It should not add significantly to the power requirement to operate the machine, particularly at the start.

Types and Characteristics of Greases

Composition
Lubricating greases consist of three components: base oil, thickener, and additives. To the base oil, a thickener and other substances are added. The properties of greases are determined mainly by the kind of base oil used, the thickening agent, and the various substances added to it.

What Does Base Oil Do in Lubricating Greases?
Normally, lubricating greases consist of 85 percent base oil, 10 percent thickener, and 5 percent performance additives. Both base oils and thickeners influence the properties of lubricating greases. Other substances are added to boost certain desired performance properties. The ability to be pumped and the flowability of lubricating greases are influenced by the properties of the base oil (see Table 3.1). The base oil can be a vegetable, petroleum, or synthetic liquid product and makes a most important contribution in terms of structure, performance, and stability.

TABLE 3.1   Base Stocks of Greases

Category	Type
Mineral oils	Paraffinic and naphthenic
Synthetic oils	Polyalphaolefin (PAO), ester, polyalkylene glycol (PAG), and alkylbenzenes
Natural oils	Vegetable oils

Greases based on high-viscosity oils flow slowly compared with greases based on low-viscosity oil. Therefore, greases based on high-viscosity oils are used for low-speed applications and where bearing diameters are comparatively larger. By contrast, in applications where bearings run under high speed, greases based on low-viscosity oils are recommended. The choice of a viscosity depends on intended service, for machines with higher DN factors (where D is diameter and N is the revolutions per minute), low viscosity is preferred, whereas for higher loads and low DN factors, heavier oils are more suitable.

What Does Thickener Do in Lubricating Greases?

Thickeners act as carriers of the oil and, in addition, act like sponges when oil is released at the point of application. Many of the important properties and performance characteristics of a fully formulated grease come from the actual thickener system. High-temperature capabilities are a function of the thickener, although base oil also plays some role. The water resistance characteristics of greases are also controlled by type of thickener. For example, soda-based greases are poor at water resistance, whereas aluminum-complex and sulfonate-based greases are known for their superior water resistance properties. Thickeners are partially soluble in lubricating fluid and impart a semisolid consistency to the grease.

Types of Thickeners

Thickeners are classified into two major families: soap and nonsoap thickeners. Normally, more than 90 percent of the thickners used worldwide are soap based. *Saponification* is the process used to develop soap-based grease thickeners, in which fat or fatty acids are used to react with an alkali (metallic hydroxide) to form a chemical soap. The fat can be of animal, vegetable, or synthetic origin, and the alkali is a substance of basic properties. The metallic hydroxide is the component by which the grease is classified (e.g., aluminum, barium calcium, lithium, sodium, etc.).

Depending on the types of fatty acids used, greases are classified as simple or complex greases. The grease classification depends on the type of thickener, as shown in Figure 3.2.

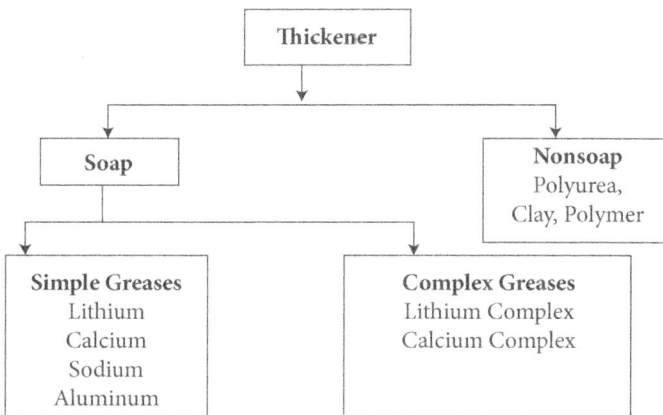

FIGURE 3.2   Classification of greases based on thickener.

Soap-Based Greases.   The main thickener used in this type of grease is a metallic soap. The metals used as thickners are lithium, aluminum, sodium, and calcium. Greases with complex soap thickeners are becoming more popular today because of their higher operating temperatures and load-carrying capabilities. Complex greases are produced by combining the metallic soap with a complexing agent. Lithium complex grease is used most widely and is made with a conventional lithium soap and low-molecular-weight organic acid as the complexing agent.

There are three types of soap-based greases:

- **Simple soaps.** Simple-soap greases are produced by the reaction of fatty acid, such as 12-hydroxystearic acid (12-HSA), with a metallic hydroxide, such as lithium hydroxide. This results in a simple lithium soap that is used most commonly worldwide. The thickener type is defined by the metallic hydroxide used. For example, if calcium hydroxide were used with a fatty acid, the grease would be called a simple calcium soap.
- **Mixed soaps.** Mixed-soap greases are produced by reaction of a fatty acid with two metallic hydroxides. For example, if 12-HSA reacts with lithium and calcium hydroxide, it would produce a mixed Ca/Li soap.
- **Complex soaps.** These types of greases are produced by mutual crystallization of two or more substances, resulting in improving the functional characteristics of the greases and higher dropping points. Say, for example, that a fatty acid such as 12-HSA is reacted with a short-chain complexing acid such as azelaic acid, the result is a complex soap. If lithium hydroxide were used, the result would be lithium complex grease. The dropping point of lithium complex greases is higher than that of simple lithium soap greases because of the presence of a second thickener component, known as the *complexing agent*. The advantage of this type of thickener over a simple soap is that it results in much better high-temperature properties.

Nonsoap Greases.   Nonsoap thickeners are used in special applications such as bentonite clay for high-temperature applications where it does not melt.

What Do Additives Do in Lubricating Greases?

Various substances are added to grease to improve certain properties and efficiency. Some additives are needed to augment or improve performance, and some are used to meet special needs. Some additives modify the soaps; others enhance the natural characteristics of the oil, give it longer life, or improve its ability to protect equipment. The additives doped in the formulation may either add new characteristics or boost desired characteristics that are already present (see Table 3.2).

TABLE 3.2   Grease Additives and Functions

Additive	Function
Antioxidants	Retard oxidation of base stock for longer lubricant life
Rust inhibitors	Protect ferrous surfaces from rusting
Antiwear agents	Provide wear protection during boundary lubrication
Tackifiers	Enhance water resistance and metal adhesiveness

Types and Characteristics of Grease Components

A detailed description of grease base oils, thickening agents, and additives is provided in the following subsections.

Base Oils

The properties of any grease are influenced mainly by the properties of the base oil. The base stock can be vegetable, petroleum, or synthetic. Generally, greases with low-viscosity base oils are best suited for low-temperature and high-speed applications; greases using high-viscosity base oils can be used for high temperatures and high loads. The different types of base oil characteristics are described next.

Petroleum Oils.   Petroleum oils used as lubricants in grease vary widely in type. The two main classes of petroleum base oils, paraffinic and naphthenic, have different effects on thickeners. Because of the relative stability of paraffinic oils, they are less likely to react chemically during grease formation. Paraffinic oils are poorer solvents for many additives used in greases. Naphthenic oils, particularly when some unsaturates are present, can function chemically during manufacture. Naphthenic oils are most common despite their relatively low viscosity index. Their low-temperature fluidity and ability to combine readily with soaps contribute to their wide use. Viscosity, viscosity index, and chemical characteristics are each important for base oil. A base oil viscosity range of 65–175 cSt at 40°C (~300–800 Saybolt Universal Seconds [SUS] at 100°F) is most common. Greases for low-temperature or high-speed use may have lower-viscosity base oils, whereas greases used specifically for low speeds, high loads, and shock loading will be higher in viscosity.

Synthetic Fluids.   These have proven to be particularly well suited for extreme conditions. Among soap-type greases, more synthetic fluids are thickened with lithium soap than with any other. Synthetic fluid greases are normally designed for improved performance in some extreme temperature range, either high or low. Products of this type find their greatest application in high-performance aircraft, missiles, and space vehicles. When thickeners and fluids are both synthetic, use is almost exclusively in such high-performance equipment.

Vegetable Oils.   Greases based entirely on vegetable oils can be formed by mixing preformed soap and lubricant in the required proportions to form a grease with the desired properties. Alternatively, grease can be formed from the same vegetable oil by simultaneous alcoholysis and saponification of the oil to form the lubricant and soap. Total vegetable oil greases have the advantage of good biodegradability, and they conform to the requirements for high performance.

Thickening Agents

To maintain the semisolid state of a grease, thickening agents are compounded with base oils. These agents control many of the important properties and performance characteristics of the grease. A grease thickener can influence the operating temperature and load, speed, and material compatibility of a specific grease. Thus it is necessary to determine the right thickener for use in specific applications. Many special characteristics of greases, such as limiting temperature range, mechanical stability, and water resistance, largely depend on the type of thickening agent used. The different types of thickeners are described below.

Simple Soaps.   The earliest greases were made when lime reacts with vegetable oil or animal fat in the presence of water to produce a *calcium soap* of the natural fatty acid. These greases were used for simple lubrication of cartwheel and waterwheel shafts and bearings.

   Sodium Soaps. Similar to domestic soap used for washing, sodium soaps have higher melting points than calcium soaps. Greases made with sodium soaps are resistant to rust and oxidation at high temperatures. Sodium soap greases were used for early machinery of the industrial revolution, and they are commonly used for lubrication of steam engines. They have an operating capability in temperatures up to approximately 110°C, although they have melting points in the range of 150–230°C. Because they have the ability to absorb minor amounts of water, they are used where rust preventive properties are required. Sodium soap greases are water soluble, and thus like washing soap, they are readily washed away by large amounts of water. They also suffer from hardening in storage. Because of the large fiber size of sodium soaps, they contribute less lubricity to the grease and, as a result, have poorer load-carrying capabilities. Hence a base oil of higher viscosity is needed to provide heavy-duty properties. Sodium soap greases also generally lack the oxidation resistance of lithium and clay greases.

   Aluminum Soaps. Greases made with aluminum soaps were used for improved lubrication for steam engines. Aluminum thickeners in the grease offer both water tolerance and a higher temperature capability. Aluminum soap greases have a smooth texture and are water resistant. But aluminum-thickened greases have a major weakness in that they are extremely sensitive to shear. They are easily broken down if load is applied, and as a result, they lose their consistency and lubricating capability. When they are exposed to temperatures above about 77°C (170°F), the normal smooth structure becomes rubbery. In the rubber-like state, the grease pulls away from metal and ceases to lubricate. These greases also tend to aerate badly when severely agitated or churned. Aluminum stearate grease can still find applications in low-shear simple plain bearings and as chassis grease. Aluminum complex grease performs as well as lithium complex grease in many situations, but it is more expensive to produce. It takes quite a bit longer to manufacture aluminum complex grease.

Calcium Soaps. Mineral oils thickened with calcium soap were among the first lubricating greases. In the manufacture of calcium soap greases, the reaction of animal fat and lime to make soap typically takes place at temperatures far above the boiling point of water. Because calcium soap greases are water resistant, they find many uses in food plants, water pumps, wet industrial and sewage plant machinery, equipment exposed to weather, marine hardware, and chassis lubrication.

Lithium Soaps. Barium- or lithium-based greases have the desirable combination of both high melting points and resistance to emulsification by water. However, they have the disadvantages of expensive raw materials, difficulty of manufacture, and/or high soap base content for any given grease consistency grade. Nowadays, almost all lithium greases are based on the castor oil derivative 12-HAS, although early lithium soaps were made from simple stearic acid derived principally from beef tallow. Lithium soaps are the most versatile and widely used greases. They are buttery and have a dropping point of around 180–190°C (up to 210°C for extreme applications). Lithium soap greases are resistant to water and oxidation and have relatively good high- and low-temperature characteristics and good mechanical stability. They are widely used as multipurpose greases and are particularly suited for rolling-element bearings.

Barium Soaps. Barium soap greases were among the first multipurpose products with both high-temperature capability and good water resistance. Impurities can affect the structure and performance of these greases. Typically, soap content to achieve a given consistency is very high. Barium soap greases are frequently used to lubricate electrical cables for power transmission lines. Wind, temperature, and current surges cause flexing and stretching between the conductors and the sheath.

Complex Soaps.   Complex greases are formed when derivatives of a single metal react with a combination of different types of acids that can be crystallized into the same fibrous thickener structure. For example, calcium complex soap greases may be formed from calcium 12-HSA and calcium acetate. The principal advantage of most complex soaps is their higher dropping points. This permits their use in applications in which the temperature may at times exceed the melting point of simple soaps and thus is the reason why they are much used today.

Depending on the specific grease application, complex greases are manufactured with high-quality naphthenic, paraffinic, or mixed naphthenic-paraffinic oils in different viscosity grades (VGs). Calcium, lithium, and aluminum complex soap greases are fairly common, with certain other types being used occasionally.

Nonsoap Thickeners.   Several nonsoap thickeners have been developed over the years and are used in applications depending on the operating conditions the grease will experience and the materials with which it will be in contact. Some thickeners need to be nonmelting and withstand temperatures significantly hotter than what traditional soap greases can handle. Others need to be inert because of material compatibility, or they may be in aggressive environments where they are exposed to corrosive or acidic environments.

In nonsoap greases, inorganic, organic, and synthetic materials are used as thickeners.

Clay Thickeners. These thickeners are generally bentonite or hectorite clay that has been chemically treated to make it thicken oil. Bentonite clays have been used in grease formulations since the beginning of the twentieth century to improve high-temperature performance. The chief feature of clay-thickened greases is that the thickener does not melt; hence these greases can be used in operations in which temperatures occasionally exceed the melting points of other thickeners. Their oxidation stability is generally no better than that of other petroleum products. Therefore, if these greases are used in sustained service at temperatures above about 120°C, frequent relubrication is necessary unless the product has been explicitly formulated for sustained service at higher temperatures. The stability of clay-based greases is limited because of their lack of a fibrous matrix structure. Sometimes residue of abrasive clay is deposited on machine surfaces owing to oil oxidation and separation.

Polytetrafluoroethylene (PTFE). These types of greases are normally used for contact with aggressive solvents and strong acids and alkalis. PTFE greases operate well under low pressure, such as in vacuum pumps, and in high-speed bearings in vacuum environments (space). A drawback to PTFE is that it is an inefficient thickener and typically can require up to 50 percent thickener depending on the stiffness of the resulting grease. PTFE can also be used to thicken other base oil chemistries such as PAOs and silicones, which are used in automotive and damping applications. PTFE greases are generally considered to be high-temperature greases that have good thermal stability, water resistance, shear stability, and lubricity. PTFE greases typically have dropping points over 260°C.

Polyurea. Polyurea greases are also characterized by dropping points in excess of 250°C and low oil separation. One of the important properties of a grease with a polyurea thickener is its excellent antioxidant capability. These greases exhibit extremely good high-temperature performance and have, in many cases, become the preferred choice for filled-for-life applications in both bearings and joints. Drawbacks of this type of grease are poorer performance at ambient temperatures and the raw materials are of a toxic nature.

Additives
Additives play several roles in lubricating greases. These include enhancing the existing desirable properties, suppressing the existing undesirable properties, and imparting new properties. Common additives used in greases are oxidation and rust inhibitors, antiwear agents, and friction-reducing agents.

Antioxidants.   Antioxidants are additives used to prolong the oxidative resistance of the base oil and, as a result, enhance the life of a lubricant. These oxidation inhibitors are the most common additives and must be selected to match the specific grease. The main objective of this additive is to protect the grease during storage prior to use. Antioxidants also allow lubricants to operate at higher temperatures than would otherwise be possible without them. Many lubricants, especially hydrocarbon-based lubricating oils, are susceptible to degradation by oxygen. Most multipurpose greases designed for high-temperature applications contain oxidation inhibitors to ensure extended service life and enable longer regreasing intervals.

Rust and Corrosion Inhibitors.   Rust is a form of corrosion formed by electrochemical interaction between iron and atmospheric oxygen and accelerated in the presence of moisture owing to the catalytic action of water. The electrochemical oxidation of the surface of iron or steel can be prevented by adding specific water-blocking substances to lubricating greases that inhibit the formation of rust. Rust inhibitors incorporated into lubricating grease provide a protective film against the effect of moisture, water, and air. Corrosion inhibitors work by neutralizing corrosive acids formed by the degradation of base fluid and lubricant additives.

Under extremely wet or corrosive conditions, the performance of most greases can be improved by a rust inhibitor. Simple sodium grease will give rust protection where traces of water are present. In industrial applications, such as food preparation and handling, suitable inhibition is a necessity. Most high-quality greases recommended for multipurpose uses contain rust and corrosion inhibitors.

Tackiness Additives.   Grease may be formulated to withstand the substantial impact common in heavy-equipment applications. The adhesive and cohesive properties of grease can be improved to resist throw-off from bearings and fittings while providing extra cushioning to reduce shock and noise via the use of tackiness additives. High-molecular-weight polymers polyisobutelene, polybutene, and latex compounds are typical examples of tackiness additives.

Extreme Pressure (EP) Additives.   When high temperature and high pressure occur during boundary lubrication, EP additives chemically react with the metal surface to be protected, forming a sacrificial coating that prevents the two metal surfaces from welding together. These EP additives produce a surface that is softer than the unprotected base metal. EP additives provide improved load-carrying ability and give added protection under shock loads. In tapered roller bearings, where thrust loads are high, EP additives are often needed to prevent scoring and wear of the roller ends. In other extreme pressure conditions, the EP agents react with steel surfaces to form a surface or interface that acts like a solid lubricant and prevents metal-to-metal contact or welding. EP additives for greases generally contain various combinations of sulfur and phosphorus. They are similar, if not identical, to those used in industrial oils and gear lubes.

Classification of Greases: NLGI Grease Grade Numbers

The National Lubricating Grease Institute (NLGI) has standardized a numerical scale for classifying the consistency of greases on the basis of American Society for Testing and Materials (ASTM) worked penetrations. The NLGI grades and corresponding penetration ranges, in order of increasing hardness, are shown in Table 3.3. NLGI has established consistency or grade numbers ranging from 000 to 6 corresponding to specified ranges of penetration. The table lists the NLGI grease classification along with a description of the consistency of each classification.

TABLE 3.3   NLGI Grease Classification Chart

NLGI Number	ASTM Worked Penetration 0.1 mm (3.28 ft) at 25°C (77°F)	Consistency
000	445–475	Semifluid
00	400–430	Seimfluid
0	355–385	Very Soft
1	310–340	Soft
2	265–295	Common Grease
3	220–250	Semihard
4	175–205	Hard
5	130–160	Very Hard
6	85–115	Solid

As Table 3.3 indicates, the grading system separates the lubricants into two groups and then classifies them into two semifluid grades and from soft to hard in six grades. *Common grease* is NLGI Grade 2. It is normally used for ordinary grease guns. *Thin greases*, which are below NLGI Grade 2, pump more easily in central lubricating systems and generally allow operation at lower temperatures. They are sometimes used in cases of high speeds (>5,000 rpm). *Heavy greases*, which are above NLGI Grade 2, have limited use. A classic use of Grade 6 block grease is in a feed box for journal bearings in old paper mills. Block, or sock, greases were once used for railroad journal wheel bearings. Heavy greases are sometimes used for high-speed bearings above 5,000 rpm. The high-speed theory is that the heavy-grease channels allow the bearing to rotate at a high rate of speed, but a trace amount of grease contacts the bearing, keeping it lubricated.

Grades 000 and 00 are used most frequently in relatively low-speed gearboxes, usually larger ones with less effective seals. These grades are not appropriate for grease-lubricated antifriction bearings because they would leak. Antifriction bearings are most frequently lubricated with Grade 2 grease because of ideal oil release and good feeding ability.

Grades 0 and 1 are most suited for central lube systems with lengthy tubing runs. Grade 3 grease may be better for large rolling-element bearings that hold substantial quantities of grease. Harder greases may be used on large, open gears or large shaft bushings in which a block of grease is resting on the rotating element. Greases that are harder than Grade 3 constitute a very minor proportion of lubricating greases.

A harder grade such as NLGI Grade 4 would be preferred in water pumps; a softer, more easily pumpable grease would be used to lubricate through fittings of farm equipment, some industrial applications, and other light-duty specialized requirements such as tractor track rollers, mine cars, and textile machinery. For many of these applications, calcium greases are being replaced by more versatile multipurpose products.

Properties of Greases

The performance characteristics used in selecting and specifying greases are described next.

Consistency and Penetration

The most important feature of grease is its consistency or rigidity. Grease consistency is its resistance to deformation by an applied force. It is a measure of the relative hardness or softness and may indicate something of flow and dispensing properties. Grease that is too fluid may leak out, and grease that is too stiff may not feed into areas requiring lubrication. Grease consistency mainly depends on the viscosity of its base oil and the type and amount of thickener used. NLGI has classified the consistency of greases by the *cone-penetration test* on the basis of ASTM D 217.

Apparent Viscosity.   Grease has a resistance to motion at startup owing to high viscosity. As grease is sheared between wearing surfaces, its resistance to flow decreases, and it moves faster. As the rate of shear increases, its viscosity decreases. In contrast, oil at constant temperature would have the same viscosity at startup as it has when it is moving.

The apparent viscosity of grease is measured according to ASTM D 1092. This measurement is useful because apparent viscosity versus shear can be helpful in predicting pressure drops in a grease distribution system under steady-state flow conditions at constant temperature.

In industrial applications, apparent viscosity is useful in predicting:

- How grease might actually perform in a bearing
- Leakage tendency from a journal bearing
- Performance at low temperatures
- Pumpability and flowability

Bleeding (ASTM D 1742).   When the liquid lubricant separates from the thickener, it is called *bleeding*. Bleeding is induced by long storage periods and exposure to high temperatures. Practically all greases will separate oil under certain conditions. In most applications, the separation of a limited amount of oil is not harmful. In some cases, separation permits lubricant to creep into narrow clearances by capillary action, where soap won't go. Pressure and vibration promote separation; this may cause trouble with soap plugging central lube systems and pressure grease cups.

Dropping Point.   As grease temperature rises, the penetration increases, and the desired consistency is lost. The temperature at which a grease becomes fluid enough to drip is called the *dropping point*. The dropping point is an indicator of the high-temperature resistance of grease. The dropping point is the upper temperature limit at which grease retains its structure, not the maximum temperature at which grease may be used. Sometimes a few types of grease regain their original structure after cooling down from the dropping point. The limiting temperature for prolonged exposure is well below the dropping point. This is particularly true of products containing volatile components or additives. ASTM D 2265-06 (2014) Standard Test Method for Dropping Point is used for measuring the dropping point of lubricating grease over a wide temperature range.

Grease Selection

The following parameters are necessary for engineers or maintenance personnel to select the best grease for a particular job.

Temperature.   The grease dropping point should be about 27°C (50°F) above the bearing temperature so that the grease does not liquefy. The evaporation rate of the base stock at various high temperatures can be used as a parameter for choosing grease. If the evaporation rate is high, then the grease may be unsuitable or may require many applications.

Evaporation should be low (5 to 10 percent maximum) at the use temperature. The higher the evaporation loss, the more frequently grease will need to be replaced. Evaporation that is too fast will cause grease hardening because of a lack of oil and increased concentration of thickener.

As an indicator for low-temperature applications, the pour point of the grease base stock is used as the minimum temperature at which the grease will not be frozen. Thinner greases, such as NLGI Grades 0 or 00, pump more easily at low temperatures than does NLGI Grade 2 grease.

Speed.   Using conventional greases, the speed (rpm) of a bearing or gear is usually not a problem. Most greases can easily handle 3,500 rpm. The normal speed of electric motors is 1,750 rpm. Normal industrial equipment runs at less than 1,750 rpm.

The choice of grease can also be aided by using a *DN* value. This value, also called the *speed factor*, is the product of multiplying the diameter of a bearing by the number of revolutions per minute, that is, *DN* = diameter *D* (in mm) × *N* (rpm). The shaft diameter should be used for journal (sleeve) bearings and the pitch diameter for antifriction bearings. Table 3.4 identifies grease selection based on *DN* values.

TABLE 3.4   Selection of NLGI Grade Grease at Various *DN* Values

DN Value	NLGI Grade
0–50,000	3
50,000–125,000	2
125,000–250,000	1
250,000–350,000	0

Pressure.   Although most industrial equipment does not require extreme-pressure greases, it is advisable to use greases that contain these additives. They will protect the bearings or gears when either extreme-pressure or shock-loading conditions arise.

Viscosity, Speed, and Pressure.   The relationship of these parameters is of interest as a point of reference. The base oil viscosity of grease is the primary determining factor in its ability to provide a proper lubricating film. The relationship that exists for the oil portion of the grease is coefficient of friction = ZN/P, where Z is the viscosity of the oil, N is the speed in revolutions per minute, and P is the pressure. To understand this formula, consider that most high-speed operations are associated with low-viscosity oil and light loads and that most low-speed operations are associated with high-viscosity oil and heavy loads. One must choose the minimum viscosity base oil required in grease to adequately lubricate the bearing at the operating temperature.

Conditions, Environment, and Contamination.   Most industrial greases are used inside plants with clean, controlled conditions. Grease selection is an issue when it is applied in unusual environmental conditions.

Aluminum complex grease should be used where large quantities of water run over the bearing, and calcium complex grease is effective for more incidental water contact. Where coal dust and paper particles are predominant or in a foundry atmosphere, lithium complex grease or grease with good flow characteristics (NLGI Grades 1 or 2) and low-viscosity base stock are used to allow dust to wash out. Grease made from chlorofluorocarbon base stock are used for harsh chemicals, such as chlorine and oxygen. Calcium sulfonate is particularly good in the presence of acids.

Pumpability is an important property because grease must reach the bearing or gear that needs lubrication. Because grease guns develop extremely high pressures, they can pump almost any grease directly into a bearing. Small-diameter lines between the grease gun and a bearing can be a problem. The grease must be carefully selected when small lines and centralized grease systems are used. Greases made with high-viscosity base stocks are harder to pump than those made with thin base stocks.

Incompatibility of Greases

The compatibility of greases is largely controlled by the thickener type. If the thickeners of two greases are the same, then they will probably be compatible. Greases made with different thickeners are generally incompatible because of the chemical reaction between the thickeners. Compatibility problems arise primarily when grease is supplied through a gun into a jerk fitting and then into the bearing. To avoid problems, the grease gun should be slowly pumped until the grease being exhausted from the bearing is the same color as the grease being pumped. This purging procedure normally replaces about 90 percent of the old grease. If only about 50 percent of the grease is replaced, then the grease mix usually liquefies and runs out of the bearing in 5–10 minutes.

Oil Versus Grease

Oil lubrication may be used in applications where normal operating temperatures are high as a result of excess heat generated by the machine or an external heat source. Grease is used when it is not practical or convenient to use oil. The choice of lubricant for a specific application is determined by operating conditions, machinery design, frequency of relubrication, maximum temperature, and environmental conditions and desired lubricant characteristics. A temperature rise resulting from friction in a bearing is generally lower with grease than with an oil bath, provided that the appropriate type and amount of grease is used and that it is supplied to the bearing in a suitable manner. Oil lubrication should be used when the relubrication interval for grease is too short.

LIQUID LUBRICANTS

Liquid lubricants are used in much larger quantities in industry and transportation over solid lubricants because they have several advantages. The most important advantage of liquid lubricants is the formation of a hydrodynamic film. Liquid lubricants are classified based on the origin of the liquid. They can be classified as mineral oil, vegetable oil (e.g., castor and rapeseed oils), animal oil (fish oil), or synthetic lubricant.

Classification of Liquid Lubricants

- **Vegetable (castor, rapeseed) oils.** These oils are less stable (rapid oxidation) than mineral oils at high temperatures, and they contain more natural boundary lubricants than mineral oils.
- **Animal oils.** These oils exhibit extreme pressure properties, but availability is a problem.
- **Mineral oils.** These oils are the most popular and most commonly used, and they are readily available.
- **Synthetic lubricant.** Viscosity does not vary as much with temperature with these oils as in mineral oil, and the rate of oxidation is much slower. Cost is high, however.

The composition, advantages, and disadvantages of the different types of oil are described next.

Vegetable Oils

Lubricants from vegetable base stocks are among the oldest lubricants known to humans. They were widely used in the industrial revolution for lubricating steam engines, textile equipment, and many other applications. Until petroleum-based products became available, vegetable-based materials were used as lubricants, and they provided excellent lubricity and extreme pressure characteristics with low toxicity. Until the end of World War I, aviation and other high-performance racing engines used castor oil in their crankcases because of its exceptional lubricity.

Vegetable oil appears to be economical but is much costlier because it does take time and money for processing. Vegetable oils are not as economical as mineral oils, and they are less stable in terms of thermal and oxidation characteristics. At higher temperatures, they start deteriorating and become oxidized when heated, meaning that the viscosity will increase.

More important is the reduction in environmental damage should the material find its way into the environment. In some areas, governments have mandated the use of fully biodegradable lubricants in delicate ecosystems.

The main advantage of vegetable oils is that they provide very good natural lubricity and are easily biodegradable. This is advantageous in areas sensitive to lubricant contamination. The main disadvantages of these natural oils are availability and their tendency to form waxes at low temperatures. Also, their oxidative stability is not as good as that of petroleum-based oils.

Mineral Oils

Currently, the most common liquid lubricants are mineral oils, which are made from petroleum. Mineral oils are widely used because they are available at relatively low cost (in comparison with synthetic lubricants). The commercial mineral oils are various base oils (comprising various hydrocarbons) blended to obtain the desired properties. In addition, they contain many additives to improve performance, such as oxidation inhibitors, rust-prevention additives, antifoaming agents, and high-pressure agents.

Mineral oils are blends of base oils with many different additives to improve the lubrication characteristics. The base oil components are compounds of hydrogen and carbon referred to as *hydrocarbon compounds*. The most common types are paraffin and naphthenic compounds.

Paraffinic Components.   Paraffinic oils are the most widely used type of lubricating oil base stock. Paraffinic oils are high in paraffin hydrocarbons and contain some wax. The paraffin components determine the pour point of the oil. The straight-chain paraffin of high-molecular-weight oils raises the pour point of oils (waxy compounds) and should be removed by dewaxing processes. Paraffinic oils are more resistant to oxidation and have lower volatility. They have a higher viscosity index and generally are better lubricants than naphthenic oils. Because naphthenic oils are normally wax free, they have low pour points and good viscosity/temperature characteristics. In general, paraffinic components are reasonably resistant to oxidation and have a particularly good response to oxidation inhibitors. In general, their viscosity index will range from 85 to 100.

Advantages of Paraffinic Oils. These oils have a high viscosity index (VI = 90–115), high flash points, good film strength, and good oxidation stability, good thermal stability, low volatility, and pour points higher than those of naphthenic and aromatic oils.

Disadvantages. The main disadvantages are high wax content, carbon residue (varnish), low wettability to metal, and high pour points.

Uses. Paraffinic oils are mainly used for gear oils, hydraulic oils, engine oils, and circulating oils.

Naphthenic Components.   Naphthenic oils have a higher content of cyclic hydrocarbons such as naphthalene and cyclohexane. They have higher density and viscosity than the paraffinic oils. They have pour points from −50 to −12°C and a larger change in viscosity with an increase in temperature. In general, their viscosity index will range from 0 to 60. But they have inferior viscosity/temperature characteristics. Naphthenic oils have low pour points over the paraffinic oils; hence they do not contribute to wax formation. Naphthenic components tend to have better solvency power for additives than paraffinic components, but their stability to oxidative processes is inferior.

Naphthenic oils are high in naphthenic hydrocarbons and contain little wax. In applications that operate over a wide range of temperatures, naphthenic oils generally would be less suitable than paraffinic oils. Naphthenic products are usually used in applications exhibiting a limited range of operating temperatures and requiring a relatively low pour point. In addition, naphthenic oils tend to swell seal materials more than most paraffinic oils.

Both naphthenic and paraffinic oils have a wide range of flash and pour points.

The main advantages of naphthenic base oils are good solubility of additives, good detergency, and low carbon residue, and a lower pour point than paraffinic oils (and therefore good for low-temperature applications).

The main disadvantages are a low viscosity index, lower oxidation stability, and lower flash points than paraffinic oils. At higher temperatures, oxidation leads to undesirable sludge type deposits.

Uses. Naphthenic oils are mainly used for general-purpose lubrication, chain oils, and pneumatic oils.

Aromatic Components.   Carbons in aromatic petroleum stocks are joined in hexagonal geometric structures with alternating double bonds between pairs of carbons. Often two aromatic rings may be fused together. Use of aromatics is falling into disfavor because of their toxicity. Aromatics can be thin oil (100-second), medium oil (30 weight), or heavy oil (brightstock, 140 weight).

Disadvantages of Mineral Oils

1. Tendency to varnish
2. Lower viscosity index (which necessitates changing oils and fluids with varying environmental and operating conditions)
3. Not as biodegradable as natural products or some synthetics
4. Can contain hazardous ingredients not removed in the refining process

Synthetic Lubricants

A trend toward the use of synthetic lubricants began during the period from 1920 to 1945. The need arose for various reasons. One reason was the increasing shortages of petroleum supplies in countries such as Germany, France, and Japan. Another reason was the demand for lubricants for new machines that required operability over greater temperature ranges than could be satisfied by petroleum.

Synthetic-based lubricants are produced to provide a product with precise and predictable properties through the chemical reaction of materials of a specific chemical composition. These oils are the result of reacting lower-molecular-weight compounds and turning them into larger molecules with desired properties. These lubricants have the potential of satisfying a wide range of requirements because they can be formulated with nearly any desired range of a specific property. However, certain other properties fixed by the chemical structures must be accepted in many cases.

Generally, synthetics have good thermal and oxidation stability, a broader temperature range, good chemical resistance, high service life, and a high viscosity index. The advantages of using synthetic lubricants become evident under conditions of high speeds and extremes of temperature. The main disadvantage of synthetic lubricants is their limited lubricity and higher cost.

Synthetic lubricants are superior to petroleum lubricants in most circumstances. Despite the superior performance, their use is limited to specific applications only because of their cost.

Gaseous Lubricants

Gaseous lubricants are among the simplest, lowest-viscosity lubricants known and include air, nitrogen, oxygen, and helium. They are applied in aerodynamic and aerostatic bearings. Because the chemical properties and aggregate state of most gases remain unchanged over a wide temperature range, gaseous lubricants offer several advantages over liquid lubricants. First, they can be applied at both very high and very low temperatures. Their chemical stability eliminates any risk of contamination of the bearing by the lubricant, and this is important for the machinery used in many branches of industry, particularly in the food, pharmaceutical, and electronics industries.

A useful property of gases is that their viscosities increase with temperature, whereas the opposite is true of liquids, resulting in a load-carrying capacity of gas-lubricated bearings increasing with temperature. However, the relatively low viscosity of gases generally limits the load-carrying capacity of self-acting aerodynamic bearings to 15–20 kPa.

Liquid Lubricant Selection

When choosing a lubricant for a particular piece of equipment, the equipment manufacturer's operation and maintenance manual should be consulted. The operation and maintenance manual will usually outline the required characteristics of the appropriate lubricants as well as a recommended schedule for replacement or filtering. If the maintenance manual is not available or is vague in its recommendations, lubricant manufacturers and distributors are other sources. Several factors need to be considered in the selection of the proper lubricant. The most important things you need to know are:

1. Operating speed
2. Load (type and magnitude)
3. Bearing type (if applicable)
4. Temperature
5. Method of lubrication
6. Operating environment

All the pertinent information on the equipment and unusual operating conditions should be provided to the lubricant manufacturer or distributor so that a lubricant with the proper characteristics can be chosen. Whenever possible, lubricants should be purchased that can be used in several applications. By limiting the number of lubricants onsite, the chance of mixing different lubricants or using the wrong lubricant is minimized.

Operating Speed

Speed as it relates to lubrication is actually composed of both the revolutions per minute and the size of the bearing or moving part. Rollers in a 5-inch bearing operating at 1,200 rpm are moving a lot faster than those in a 1-inch bearing at the same speed. So linear velocity, which takes into consideration both D and N, uses the following equation:

$$\text{Linear velocity} = D \times N$$

where the diameter D is the outside diameter of the inner race or cone and N is the revolutions per minute. Speeds are categorized as follows:

Slow = <4,572 cm/min (1,800 in/min)
Moderate = 4,572–25,400 cm/min (1,800–10,000 in/min)
Fast = >25,400 cm/min (10,000 in/min)

In general, greases are recommended for velocities up to 19,050 cm/min (7,500 in/min). Velocities ranging from 19,050 to 29,210 cm/min (7,500–11,500 in/min) should be bathed in oil to ensure proper lubrication, and finally, velocities in excess of 29,210 cm/min (11,500 in/min) are best lubricated with circulating oil systems, with a cooler being installed for speeds approaching 48,260 cm/min (19,000 in/min). In general, higher speeds require thinner greases and thinner base oils to prevent channeling and resulting lubricant starvation or excessive heat generation from fluid friction.

Load

Loads of 14 kg/cm² (200 lb/in²) or less are considered to be light, and loads of 35 kg/cm² (500 lb/in²) or more are extreme-duty service. Any shock loading at all or sudden stopping and starting should be considered severe service. It can be assumed that larger bearings (over 5 cm or 2 inches) are carrying heavy loads; otherwise, there would not be the need for a very large bearing. EP additive oils should be used for high-load applications.

Temperature

The operating temperature of the mechanism to be lubricated is important for two reasons. First, because the viscosity of a lubricant will go down under increasing temperatures, it is important that the one chosen retains a sufficient film to adequately reduce friction. Second, there are definite temperature limitations to different base stocks, and if exceeded, varnishes will form that can actually do more harm than good. Generally, petroleum-based lubricants should not be used above 150°C (300°F) on a regular basis. Ester-based synthetic lubricants can be used up to 230°C (450°F) and polyolesters up to 270°C (520°F). Silicones can be used up to 205°C (400°F). Other specialty synthetics such as fluorosilicones can go as high as 230°C (450°F). Synthetics tend not to char or varnish the way petroleum-based oils do, and because of this, they will not form hard deposits that can accelerate wear.

Method of Lubrication

The type of lubricating system may dictate the type of lubricant to be used. For example, a bearing enclosed in a gearbox can use oil, but the same bearing out in an open setting must have a grease to prevent lubricant loss from leakage. Higher-speed bearings out in the open may require a tacky lubricant to prevent sling-off of the applied lubricant. Automated dispensing systems are becoming increasingly common. They deliver an appropriate amount of lubricant at crucial times to extend the working life of a component. Because of the accurate metering, less lubricant is required to perform the lubrication in the long run. It is very important to use a lubricant in these systems with excellent oxidation resistance so that gummy deposits do not build up in the mechanisms.

Operating Environment

The last parameter of the tribological system is the application's operating environment. If the environment includes moisture or water, the lubricant must provide good anticorrosion properties as well as resistance to water washout or water contamination. If the application operates in a vacuum or partial vacuum, the atmospheric pressure of the application must be within the operational limit of the lubricant and above its vapor pressure at the operating temperature. If the application must operate in the presence of certain chemical liquids or vapors, the selected lubricant must be resistant to these chemicals. Even an ideal environment, such as computer room or clean-room processing facility, could have specific requirements for noise-reducing lubricants in rolling-element bearings or instrument bearings.

Physical Properties of Lubricating Oils

Viscosity

Probably the single most important characteristic of a lubricant is its viscosity. Viscosity is a measure of a fluid's internal friction or resistance to flow. It is a vital property of a lubricant because it influences the ability of the oil to form a lubricating film or to minimize friction. The higher the viscosity of a fluid, the greater is the internal resistance and the greater is its load-carrying capacity. But with the higher internal resistance, temperatures can rise. The correct viscosity for a particular application would be thick enough to support the load but not so thick as to cause excessive fluid friction and a corresponding increase in temperature. Hence the selection of proper lubricant viscosity is often a compromise between selecting one high enough to prevent metal-to-metal (wear) contact and low enough to allow sufficient heat dissipation. Viscosity varies inversely with temperature. Viscosity is also affected by pressure; higher pressure causes the viscosity to increase, and subsequently, the load-carrying capacity of the oil also increases. This property enables use of thin oils to lubricate heavy machinery. The load-carrying capacity also increases as the operating speed of the lubricated machinery is increased.

Viscosity Index.   The viscosity of a lubricant will change with temperature, increasing at low temperatures and dropping at higher ones. The viscosity index (VI) is an empirical number designed to indicate the amount of viscosity change over a given temperature range. Viscosity index can be classified as follows: low VI, <35; medium VI, 35–80; high VI, 80–110; very high VI, >110. A high VI indicates small oil viscosity changes with temperature. A low VI indicates high viscosity changes with temperature. Therefore, a fluid that has a high VI can be expected to undergo very little change in viscosity with temperature extremes and is considered to have a stable viscosity. A fluid with a low VI can be expected to undergo a significant change in viscosity as the temperature fluctuates.

Knowing the VI of an oil is crucial when selecting a lubricant for an application and is especially critical in extremely hot or cold climates.

Density

Density is important because oils may be formulated by weight but measured by volume.

Lubricity

The lubricity of a lubricant refers to how well the product reduces friction beyond what would be indicated by its viscosity alone. In other words, it indicates how well it clings to a surface once in operation. It is a property that can be enhanced by the selection of additives or by blending different base stocks. The test typically used to measure this is the Falex Pin & Vee Block Test (ASTM D 3233).

Demulsibility

Demulsibility refers to a lubricant's ability to readily separate from water. Oils used in force-feed lubrication systems should possess good water separability to prevent emulsification.

Foaming

A lubricant that is used in high-speed conditions, such as in gearboxes and pressurized lubrication systems, should have good resistance to foaming. In pressurized lubrication systems, an air leak at the inlet can cause air to be beaten in under pressure and dissolved into the lubricant at the pump. When the pressure is released, the entrained air will bubble out, resulting in foam.

Foaming is an important property because when machine elements mix air into their lubricants, the result can be:

- Reduced heat transfer
- Interference with lubricant flow
- Expulsion of lubricant through vents
- Accelerated oxidation (because of heat generated during compression)
- Spongy hydraulic system performance

Foaming is controlled by very low concentrations of antifoam additives. Additives often adversely affect air release. Foaming is measured by ASTM D 892 and other performance-type procedures.

Filterability

Filterability is the ability to remove particulate matter from lubricants by passing them through porous media. Particles of contaminants cause abrasive wear and may form deposits that interfere with lubricant flow or the motion between parts. Filterability is affected by base oil type and viscosity, additives used for other purposes, and operating conditions. It is determined by a variety of performance tests.

Neutralization Number.   As petroleum products are subjected to elevated temperatures, the process of oxidation occurs. Oxidation leads to the formation of organic acids in the lubricant. This increase in acidity reduces the water-separating ability of certain oils and may also prove corrosive to certain alloys. The neutralization number measures the amount of acidity present in the lubricant. It is quantitatively defined as the amount of potassium hydroxide (KOH) required to neutralize the acid present in one gram of sample. This quantity is also referred to as the *total acid number* (TAN). The acid number for an oil sample is indicative of the age of the oil and can be used to determine when the oil must be changed. The industry standard method for determining the acid number is found in ASTM D 974.

Total Base Number. Internal combustion engine oils are formulated with a highly alkaline (base) additive package designed to neutralize the acidic by-products of combustion. The *total base number* (TBN) is a measure of this additive package, and it may be used as an indication of when diesel engine oil should be changed.

Detergency and Dispersancy

Detergency is the property that involves the suspension of oil-insoluble materials, and dispersancy is the prevention of sludge and varnish formation. The insoluble materials can be oxidation and corrosion products, reaction products of gas-phase materials, or other materials that leak into the lubricant. Both detergency and dispersancy are provided to lubricants by means of additive molecules that consist of insoluble-material-attracting polar groups and oil-attracting groups.

Low-Temperature Properties

When a sample of oil is cooled, its viscosity increases in a predictable manner until wax crystals start to form. The matrix of wax crystals becomes sufficiently dense with further cooling to cause an apparent solidification of the oil. Although the solidified oil does not pour under the influence of gravity, it can move if sufficient force is applied. A further decrease in temperature cause more wax to form, increasing the complexity of the wax/oil matrix. Many lubricating oils have to be capable of flow at low temperatures.

Cloud Point. As petroleum products consist of a mixture of molecular components, lubricants do not exhibit sharp freezing points. Rather, as a lubricant is cooled, certain components such as paraffin wax begin to form and will begin to precipitate out and become evident in the liquid as a cloud. The temperature at which this occurs is called the *cloud point* of the lubricant.

As the temperature drops, wax crystallizes and becomes visible. Certain oils must be maintained at temperatures above the cloud point to prevent clogging of filters. A sample of oil is warmed sufficiently to be fluid and clear. It is then cooled at a specified rate. The temperature at which haziness is first observed is recorded as the cloud point (the ASTM D 2500/IP 219 test). The oil sample must be free of water because water interferes with the test.

Pour Point.   The pour point is the lowest temperature at which a lubricant can be observed to flow under specified conditions. This property is crucial for oils that must flow at low temperatures. A commonly used rule of thumb when selecting oils is to ensure that the pour point is at least 10°C (20°F) lower than the lowest anticipated ambient temperature. Oils generally contain some paraffinic wax that was not removed during refining. At low temperatures, these waxes crystallize and form three-dimensional lattices that interfere with normal oil flow. Pour point is related to viscosity because its concern is whether the oil will flow at low temperatures or just barely flow under prescribed conditions.

This is an important property of lubricant base oils. High-viscosity oils may cease to flow at low temperatures because their viscosity becomes too high rather than because of wax formation. In these cases, the pour point will be higher than the cloud point.

The pour point is important in choosing a lubricant for cold weather applications. Oils used under low-temperature conditions must have low pour points. Oils must have pour points (1) below the minimum operating temperature of the system and (2) below the minimum surrounding temperature to which the oil will be exposed. It is also important in applications such as refrigerant compressor lubrication, where the oil is subjected to low temperatures (see ASTM D 5949, Standard Test Method for Pour Point of Petroleum Products).

High-Temperature Properties
The high-temperature properties of oil are governed by distillation or boiling-range characteristics of the oil.

Volatility.   Volitility is important because it is an indication of the tendency of an oil to be lost in service by vaporization.

Flash Point and Fire Point.   As a lubricant is heated, lighter components begin to vaporize. The *flash point* is the lowest temperature to which a lubricant must be heated before its vapor, when mixed with air, will ignite but not continue to burn. The *fire point* is the temperature at which lubricant combustion will be sustained. The flash and fire points are useful in determining a lubricant's volatility and fire resistance. The flash point can be used to determine the transportation and storage temperature requirements for lubricants. Lubricant producers can also use the flash point to detect potential product contamination. A lubricant exhibiting a flash point that is significantly lower than normal will be suspected of contamination with a volatile product. Products with a flash point of less than 38°C (100°F) will usually require special precautions for safe handling. The fire point for a lubricant is usually 8–10 percent above the flash point.

Chemical Properties

Corrosivity.   Corrosivity is the tendency of a lubricant and its contaminants to chemically react with ferrous and nonferrous metals. Corrosion damages bearings and other structural elements and accelerates lubricant oxidation by catalysis. It is measured in performance tests, including many standard bench oxidation tests. Consequently, the oxidation and (nonrust) corrosion properties of a lubricant are commonly considered together. Corrosion can be reduced by additives that inhibit the oxidation process, form protective films on surfaces, or deactivate the catalytic properties of dissolved metals.

Oxidation Stability.   The chemical reaction whereby oxygen in air combines with hydrocarbon in oil is called *oxidation*. Straight mineral oils possess a certain resistance to oxidation during the early stages of service, but subsequently deterioration due to oxidation tends to accelerate. Operating conditions where oxidation can occur include:

- High temperature
- Presence of metallic wear particles
- Presence of moisture and other contaminants, such as sludge, dirt, rust, and other corrosion products
- Churning and agitation

Oxidation of straight mineral oils proceeds very slowly at room temperature; at 60°C (140°F), oxidation is still slow but significant; and above 93°C (200°F), it is greatly accelerated. Oil oxidation products are undesirable for the following reasons. Insoluble products (sludge) may prevent effective lubrication because of clogging. Soluble products circulating with the oil tend to be acidic and eventually either lead to corrosion or pitting of bearing surfaces or form varnish deposits on parts operating at high temperatures.

The significance of oxidation of a lubricant is that as oxidation progresses, the viscosity of the oil can change substantially (it usually thickens), degrading lubricating capabilities. Varnishes and sludge form that can plug oil lines and orifices.

LUBRICANT ADDITIVES

Practically all lubricants contain additives to enhance existing properties or to impart new properties. There are three general classifications of lubricant additives:

1. Surface-protective additives
2. Performance-enhancing additives
3. Lubricant-protective additives

As the names imply, surface-protective additives protect bearing surfaces, performance-enhancing additives enhance a lubricant's performance for specific applications, and lubricant-protective additives prevent deterioration of the lubricant.

Surface-Protective Additives

Lubricity Additives

Lubricity, also referred to as *oiliness*, with respect to lubricating oils is defined as the ability of an oil to reduce friction between moving surfaces. Lubricity additives, usually vegetable or animal fats, enhance lubricity by tenaciously adhering to the metal's surface, forming an adsorbed film of high lubricating value.

Antiwear Additives

The main objective of antiwear oil additives is to reduce the wear rate in sliding or rolling motion under boundary lubrication conditions. A layer of coating is developed by the anti-wear additives on the metal's surface. If there is light metal-to-metal contact between the surfaces, the heat generated from the friction melts the additives, forming a liquid layer between the surfaces. The thin layer separates the rubbing surfaces and reduces the adhesion force at the contact between asperities of the two surfaces. In addition, the molten additive, which is softer than the metal, acts as a lubricating layer over the surfaces, preventing metal wear. Addition of antiwear additives into lubricating oils helps to promote longer machine life because of higher wear resistance. Zinc dialkyldithiophosphate (ZDDP) is an effective and widely used antiwear additive.

Extreme Pressure Additives

Extreme-pressure (EP) additives prevent seizure of components caused by direct metal-to-metal contact under high loads. These additives react with the metal surface to form a compound that acts as a protective layer. Because this layer is softer than the metal, it protects it under EP conditions. As this layer is removed, the EP additive acts to form another layer. In contrast to the action of antiwear additives, EP additives control wear instead of preventing it. The following materials are used as EP additives:

- Chlorinated paraffins
- Sulfurized fats
- Esters
- Zinc dialkyldithiophosphate (ZDDP)
- Molybdenum disulfide

Tackiness Additives

An oil tackifier is an additive that holds a lubricating oil in place where it is needed. Tackiness agents act to increase the adhesiveness of an oil or grease. These are stringy materials used in some oils and greases to prevent the lubricant from flinging off the metal surface during rotational movement. Typically, tackifier is a polymer dissolved in oil that oil blenders can add to their lubricant oil. The polymer helps the oil to adhere to the metal surface of the mechanical part that it is lubricating. Lubricant blenders use poly-alpha-olefin (PAO), synthetic ester, polyisobutylene (PIB), and white oil to make food-grade lubricants.

Corrosion- and Rust-Inhibiting Additives

Chemical contaminants can be generated in an oil or enter into a lubricant from contaminated environments. Corrosive liquids often penetrate through the seals into a bearing and cause corrosion inside the bearing. This problem is particularly serious in chemical plants where there is a corrosive environment, and small amounts of organic or inorganic acids usually contaminate the lubricant and cause considerable corrosion. In addition, organic acids from the oil oxidation process can cause severe corrosion in bearings. Organic acids from oil oxidation must be neutralized; otherwise, the acids degrade the oil and cause corrosion. Oxygen reacts with mineral oils at high temperature. The oil oxidation initially forms hydroperoxides and, later, organic acids. White metal (Babbitt) bearings and the steel in rolling-element bearings are susceptible to corrosion by acids. It is important to prevent oil oxidation and contain the corrosion damage by means of corrosion inhibitors in the form of additives in the lubricant.

In addition to acids, water can penetrate through seals into the oil (particularly in water pumps) and cause severe corrosion. Water can get into the oil from the outside or by condensation. Water in the oil is a common cause for corrosion. In rolling-element bearings, the corrosion accelerates the fatigue process, referred to as *corrosion fatigue*. Rust inhibitors are oil additives that are absorbed on the surfaces of ferrous alloys in preference to water, thus preventing corrosion. Similar to rust inhibitors, corrosion inhibitors are preferentially absorbed on the surface and are effective in protecting it from corrosion. The following materials are used as rust and corrosion inhibitors:

- Alkaline compounds
- Organic acids
- Esters
- Amino-acid derivatives

Detergent and Dispersant Additives
Detergents and dispersants are used primarily in engine oils to keep surfaces free of deposits and keep contaminants dispersed in the lubricant.

Detergents.   Detergents neutralize strong acids present in the lubricant (e.g., sulfuric and nitric acids produced in internal combustion engines as a result of the combustion process) and remove the neutralization products from the metal surface. Detergents also form a film on the part surface that prevents high-temperature deposition of sludge and varnish. Phenolates, sulfonates, and phosphonates of alkaline and alkaline-earth elements such as calcium (Ca), magnesium (Mg), sodium (Na), and barium (Ba) are used as detergents in lubricants.

Dispersants.   Dispersants keep the foreign particles present in a lubricant in a dispersed form (finely divided and uniformly dispersed throughout the oil). The foreign particles are sludge and varnish, dirt, products of oxidation, water, and so on. Long-chain hydrocarbon succinimides, such as polyisobutylene succinimide, are used as dispersants in lubricants.

Friction-Modifying Additives
These are additives that usually reduce friction. The mechanism of their performance is similar to that of the rust and corrosion inhibitors in that they form durable low-resistance lubricant films via adsorption on surfaces and via association with the oil. In addition to reducing friction, the friction modifiers also reduce wear, especially at low temperatures, where the antiwear agents are inactive, and they improve fuel efficiency. Friction modifiers reduce the coefficient of friction, resulting in less fuel consumption. The crystal structure of most of friction modifiers consists of molecular platelets (layers), which may easily slide over each other. The following solid lubricants are used as friction modifiers:

- Graphite
- Molybdenum disulfide
- Boron nitride (BN)
- Tungsten disulfide (WS2)
- Polytetrafluoroethylene (PTFE)

Performance-Enhancing Additives

Viscosity Index Improvers
Viscosity index improvers lower the rate of change of viscosity with temperature. The viscosity of oils sharply decreases at high temperatures. Low viscosity causes a decrease of the

lubricating ability. These additives are usually long-chain polymeric molecules. They have a relatively high molecular weight, which is three orders of magnitude larger than that of the base-oil molecules. Viscosity index improvers keep the viscosity at acceptable levels, which provides a stable oil film even at increased temperatures. Viscosity improvers are widely used in multigrade oils, the viscosity of which is specified at both high and low temperature. Acrylate polymers are used as viscosity index improvers in lubricants.

Pour Point Depressants

The pour point is an important characteristic whenever a lubricant is applied at low temperatures. The pour point is the lowest temperature at which the oil may flow. Wax crystals formed in mineral oils at low temperatures reduce their fluidity. Pour point depressants inhibit the formation and agglomeration of wax particles, keeping the lubricant fluid at low temperatures. Copolymers of polyalkylene methacrylates are used as pour point depressant in lubricants. In addition, certain synthetic oils were developed that can be applied in a wide range of temperatures and have a relatively low pour point.

Demulsifiers

A demulsifier promotes the separation of oil and water in lubricants exposed to water. Demulsifier additives prevent the formation of a stable oil-water mixture or an emulsion by changing the interfacial tension of the oil so that water will coalesce and separate more readily from the oil. This is an important characteristic for lubricants exposed to steam or water so that free water can settle out and be easily drained off at a reservoir.

Lubricant-Protective Additives

Oxidation Inhibitors

At high temperature, oxygen reacts with mineral oils to form hydroperoxides and, later, organic acids. The oxidation process is considerably faster at elevated temperatures. in fact, the oxidation rate doubles for a nearly 10°C rise in oil temperature. Oil oxidation is undesirable because the products of oxidation are harmful chemical compounds such as organic acids that cause corrosion. In addition, the oxidation products contribute to a general deterioration of the lubricant properties. The organic acids that are products of oil oxidation cause severe corrosion of the steel journal and the alloys used as bearing materials. The oil circulates, and the corrosive lubricant can damage other parts of the machine. In addition, the oxidation products increase the viscosity of the oils as well as forming sludge and varnish on the bearing and journal surfaces. Lubricant degradation is catalyzed by the presence of metals and oxygen. It is very important to prevent or at least to slow down this undesirable process.

The oxidation inhibitors improve the lubricant's desirable characteristic of oxidation resistance in the sense that the chemical process of oxidation becomes very slow. The following materials are used as antioxidants:

- Zinc dithiophosphate (ZDP)
- Alkyl sulfides
- Aromatic sulfides
- Aromatic amines
- Hindered phenols

Foam Inhibitors

Agitation and aeration of lubricating oils occurring in certain applications (e.g., gear oils and compressor oils) may result in the formation of air bubbles in the oil—foaming.

The bubbles formed by foaming enhance oil oxidation, deteriorating the performance of hydrodynamic oil films in bearings. In addition, foaming adversely affects the oil supply of lubrication systems (it reduces the flow rate of oil pumps).The function of antifoaming additives is to increase the interfacial tension between the gas and the lubricant. In this way, the bubbles collapse, allowing the gas to escape. Dimethylsilicones (dimethylsiloxanes) are commonly used as antifoaming agents in lubricants

Additive Depletion

Some additives, such as antiwear and EP additives and rust, oxidation, and corrosion inhibitors, are consumed as they are used. When all of a particular additive has been consumed, the lubricant is no longer capable of performing as originally intended. Usually this condition requires replacement of the lubricant, but in some cases replenishment of the additive is possible. The lubricant manufacturer should be consulted before this is attempted.

CONCLUSION

Lubricants are generally composed of 90 percent base oil and less than 10 percent additives to impart desirable characteristics. Vegetable and synthetic oils are sometimes used as base oils. Additives deliver reduced friction and wear; increased viscosity and improved viscosity index; resistance to corrosion, oxidation, aging, or contamination; and so on. Dry lubricants such as graphite, molybdenum disulfide, and tungsten disulfide also offer lubrication at higher temperatures than liquid- and oil-based lubricants. Effective lubricant selection must strike a balance between quality, application, and affordability. In order to achieve and maintain this balance, lubricant specifications should be created to serve as a guideline for what to purchase and how to use it.

ICML Questions

1. Although lubricants are used to reduce friction and wear, they have some secondary functions such as
   a. cooling.
   b. sealing.
   c. cleaning.
   d. all of the above

2. Which of the following is *true*?
   a. The presence of water helps graphite in lubrication.
   b. The presence of water is detrimental to MoS_2.
   c. Vacuum is detrimental to graphite.
   d. All of the above are true.

3. Which of the following is *true* for graphite lubrication?
   a. Water vapor is a necessary component.
   b. Water vapor is detrimental.
   c. Water vapor has no impact.
   d. Water reacts with graphite.

4. Which of the following is *true* for molybdenum disulfide?
   a. It has high load-carrying capability.
   b. It can withstand high temperature and can be used in space.
   c. Moisture is detrimental to its performance.
   d. All of the above

5. Which of the following lubricants should be used in applications where low wear rate is important and temperatures go as high as 1,000°C?
   a. Polytetrafluoroethylene (PTFE)
   b. Soft metals
   c. Ceramic
   d. None of the above

6. Calcium base grease belongs to the family of
   a. soap-based greases.
   b. non-soap-based greases.
   c. Either of the two
   d. Neither of the two

7. Which of the following is *true*?
   a. Greases for low temperature or high speed use lower-viscosity base oils.
   b. Greases used for low speeds, high loads, and shock loading use base oils of higher viscosity.
   c. a and b are true.
   d. Both a and b are not true.

8. Bentonite is a
   a. clay thickener.
   b. nonclay thickener.
   c. metallic thickener.
   d. None of the above

9. Antioxidants are additives used in grease to
   a. prolong the oxidative resistance of the base oil.
   b. protect the grease during storage prior to use.
   c. operate at higher temperatures.
   d. All of the above

10. Tackiness additives are used in grease to
    a. withstand heavy impacts.
    b. resist grease throw-off from bearings and fittings.
    c. provide extra cushioning to reduce shock.
    d. All of the above

11. During a boundary lubrication regime, which of the following additives are used?
    a. Tackiness additives
    b. EP additives
    c. Antioxidants
    d. All of the above

12. Grease normally used for ordinary grease guns is
    a. NLGI Grade 000.
    b. NLGI Grade 2.
    c. NLGI Grade 1.
    d. NLGI Grade 4.

13. Grease most suited for central lube systems with lengthy tubing runs is
   a. NLGI Grade 000
   b. NLGI Grade 2.
   c. NLGI Grade 1.
   d. NLGI Grade 4.

CHAPTER 4

Pump Lubrication

Pumping systems are the single largest type of industrial equipment in process industries. Some large end users, such as chemical plants and refineries, have thousands of pumping systems. Often referred to as the "heart" of a plant, pumps are key components in a site's overall reliability program. To enhance reliability of these pumps, lubrication plays an important role. However, many process plants pay little or no attention to lubrication effectiveness. Concern for pump reliability is growing in response to increasing repair costs. As a result, professionals continue to pay more attention to lubrication issues.

American National Standards Institute (ANSI) pumps are usually smaller, overhung units for light-duty types of service. American Petroleum Institute (API) pumps (which meet API's higher standards) are typically used in heavier-duty services than ANSI pumps. ANSI pumps have both radial and thrust rolling-element bearings. Typically, the radial bearings are single-row, deep-groove ball bearings. The thrust bearings are either paired angular-contact or double-row angular-contact ball bearings. API pumps typically incorporate double-row, deep-groove ball bearings and angular-contact ball bearings for thrust loads. For larger API heavy-duty pumps, radial bearings are cylindrical—for their greater load-carrying ability—and thrust bearings are paired tapered-roller types for thrust loads.

LUBRICATION SELECTION GUIDELINES

Most pumps are lubricated with rust- and oxidation-inhibited (R&O) oils that also have small amounts of antifoam and demulsifier additives. Some pump manufacturers recommend anti-wear (AW) additives, but most use R&O oils. Viscosity is the most important criterion in the selection of a proper pump lubricant. Table 4.1 lists guidelines on proper viscosity selection for rolling-element bearings. The most commonly used viscosity for centrifugal pump rolling-element bearings is International Standards Orginization (ISO) VG 68. In some colder climates, ISO VG 32 is used.

TABLE 4.1   Viscosity Guidelines for the Lubrication of Rolling-Element Bearings

Bearing Operating Temperature	Ball/Cylindrical Bearings	Other Bearing Types
65.5°C (150°F)	VG 46	VG 68
79.5°C (175°F)	VG 68	VG 100
87.7°C (190°F)*	VG 100	VG 150

*Synthetics such as PAOs are usually recommended at this temperature.

LUBRICATION APPLICATION METHODS

The proper application of lubricants is as important as the correct lubricant selection. The choice of which lubrication method to apply for pump bearings has implications for both the short-term characteristics exhibited by the bearings and their long-term reliability. When lubricating rolling-element bearings, five basic strategies have been employed historically:

- **Oil sump.** This option establishes an oil level at the center of the bearing's bottom rolling element and represents the comparative baseline of bearing friction among the lubrication methods. Best results over time can be achieved using a constant-level oiler.
- **Oil lubrication via slinger disk or oil ring.** In this method, an oil ring is suspended from the horizontal shaft into an oil bath positioned below the bearings. The rotation of the shaft and ring flings oil from the bath into the bearings. The reduced oil volume in the bearing reduces the viscous friction in the bearing system to allow higher shaft speeds and better cooling.
- **Oil mist and air/oil.** In this case, oil is atomized and carried by an air stream to the bearing. Among all pump bearing lubrication approaches, this process generates the least amount of friction (allowing rotational speed to be based on the bearing design instead of lubrication limitations) and creates a positive pressure within the bearing housing (preventing invasive contaminants).
- **Grease.** Grease is easy to apply, can be retained within a bearing's housing, and offers extra sealing protection. Depending on the rotational speeds and operating temperatures, relubrication may be required to combat short grease lives. As an attractive alternative when the operating conditions allow, sealed "greased-for-life" bearings have been developed to eliminate the need for relubrication and related maintenance tasks.
- **Circulating lubrication method.** This method is often used on large, expensive bearings, or when it is absolutely essential to provide continuous operation under adverse conditions. In contrast to total-loss lubrication systems, after the

oil passes through the point requiring lubrication, it is fed back through the return line into the oil reservoir for reuse. In addition to lubricating, circulating-oil lubrication performs a range of other functions. It stabilizes the lubrication points to the proper temperature, removes and filters out wear particles from friction points, prevents corrosion damage, and removes water condensation. Circulating-oil systems are designed not only to lubricate but also to cool highly stressed bearings in nearly every size of machine used in the pulp and paper sector and heavy industry.

Oil Sump

Oil bath or oil sump lubrication is one of the oldest and simplest methods of oil lubrication. Here, as the shaft rotates, the rolling elements in the bearing, typically steel balls, make contact with a controlled level of oil. The most important considerations are speed, oil viscosity, and load. It is critical that an effective oil film be maintained between the rolling element and the race of the bearing. Only enough contact between the bearing and the surface of the oil as necessary to "load" the bearing with lubricant is required. If the level of lubricant is too high or too low, excessive heat will be generated, accelerating degradation of the oil and shortening the life of the bearing (see Figure 4.1). When the level of oil is too high, a condition known as *churning* occurs. Similar to the result of using an egg beater, air is "whipped" into the oil. This, along with the induced heat, increases the oxidation rate—shortening the effective life of the oil. When the oil level is too low, there is not sufficient contact to lubricate the bearing.

For smaller bearing arrangements and slower speeds, oil bath lubrication arrangements are commonly employed. In these arrangements, the normal oil level is set at around one-third to one-half the diameter of the rolling-element ball (or roller), as shown on Figure 4.1. These arrangements have the advantage of simplicity of design and manufacture.

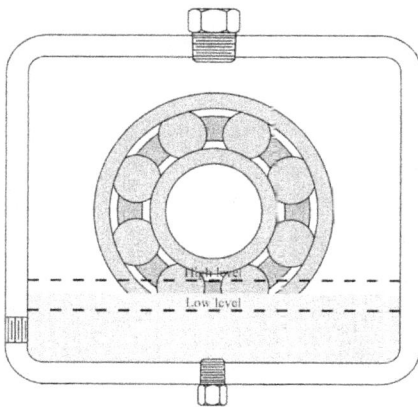

FIGURE 4.1   Oil level range in bearing housing.

Figure 4.2 shows how a bottle oiler maintains the oil level in an oil bath. As the oil level drops in the reservoir, the oiler feeds additional oil and breaks the seal in the bulb, introducing air, which allows oil to flow from the bulb. When the correct level is reached, the seal stops the oil flow.

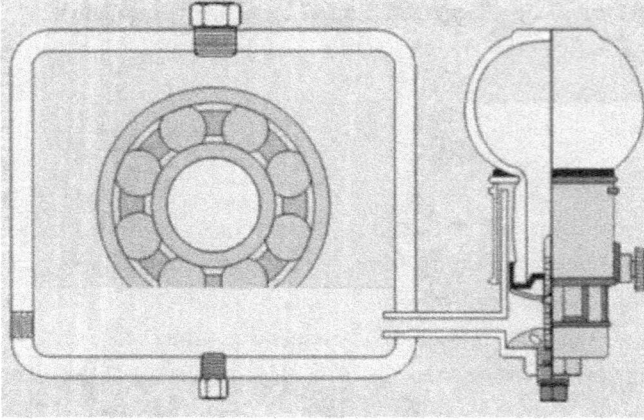

FIGURE 4.2   How a bottle oiler maintains the oil level in oil bath lubrication systems. (Source: Trico Manufacturing.)

Benefits of an Oil Sump

Oil sump lubrication is very simple by design. Understanding the basic methods and procedures can result in a very low-cost, low-maintenance, reliable system of equipment lubrication. Although simple by design, proper application may be more complex. As noted previously, it is important to understand the relationships between oil type, shaft speed, bearing housing design, and oil level maintenance. Once these factors are understood, determining the correct application is relatively easy and consistent.

Application

Perhaps the most widely used method of maintaining the proper level lubricant in a bearing housing is the constant-level oiler (Figure 4.3). Simple by design, the constant-level oiler replenishes oil lost by leakage through seals, vents, and various connections and plugs in the bearing housing. Once the proper level has been set, replacing the oil in the reservoir is the only required maintenance other than the oil changes based on preventive maintenance schedules or predictive maintenance criteria. View ports can be used to verify proper oil level.

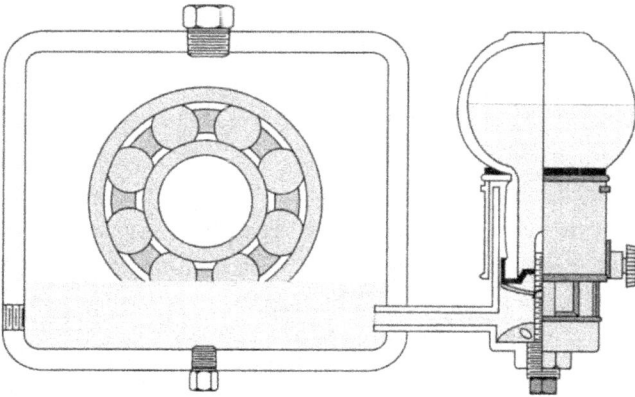

FIGURE 4.3   Constant-level oiler application.

Flinger Ring

At excessive speeds, if too much frictional heat is generated by the plowing action of rolling elements, then an oil bath method is not suitable. The oil bath lube method is avoided on process pumps whenever DN, the inches of shaft diameter D multiplied by shaft revolutions per minute N, exceeds 6,000 because heat reduces oil film strength and greatly accelerates the rate at which oil oxidizes. Bearings with DN values in excess of 6,000 will require the addition of either a flinger disk or an oil ring (Figure 4.3) or similar lube application to dependably lift or spray feed oil into the bearings.

Pumps with two-pole operation (3,000 rpm and higher) and, for larger bearings, many other manufacturers use oil lifting lubrication via an oil ring or oil slinger. The goal of these arrangements is to have the normal oil level below the rolling element bearing (hence the lifting required) in order to minimize heat generation resulting from oil churning. This churning can easily account for more than 50 percent of the total heat generated in the bearing.

Oil rings are loose pieces that rest on the shaft under gravity and the bottom point of which dips into the oil sump. As the shaft rotates, the oil ring is driven to rotate, picking up oil from the sump and depositing oil on the shaft, throwing it against the inside of the housing and atomizing it into a mist. The bearing elements make contact with the oil and undergo splash-type lubrication without direct bearing contact.

Because the oil ring is free to move, significant ranges of motion are possible, such as oscillatory (pendulum), conical, and translational motion. Additionally, if the pump installation is not level, the ring has a tendency to act under gravity and run downhill. Because of this movement, the oil ring may interact with the fixed structures in the bearing assembly in an unfavorable way, which can lead to a reduction in oil delivery together with oil ring and/or housing wear. There have been reported cases of extensive ring wear. For this reason, modern oil ring designs tend to employ guides or carriers to better control ring motion.

However, oil rings are potentially vulnerable components. They will not interact in the same way with lubricants of different viscosities or at different immersion depths. If the oil level is too high or too low, flinger ring performance is negatively affected. If the oil level is too high, the ring will become submerged, reducing its ability to lift the oil onto the rotating shaft, where centrifugal force directs oil to the bearings. If the level is too low, the ring may not be able to pick up enough oil to satisfactorily lubricate the bearings. The ring does not splash the oil directly to the bearings; it actually lifts the oil from the sump up to the shaft, where centrifugal force directs the oil to the rolling-element bearings.

Oil rings will generally malfunction if the viscosity differs much from the manufacturer's recommendations. The general catalogs of all bearing manufacturers explain that specific types of rolling-element bearings require lubricant viscosities suitable for their speed, size, and ambient temperature environment.

Unless used on perfectly horizontal shaft systems, oil rings will run downhill and then often make contact with the bearing housing. To resist deformation while operating, oil ring fabrication must include an annealing step to relieve stress. Without this annealing, many oil rings will become oval in shape. Also, oil rings tend to become progressively more unstable as DN values approach or exceed 8,000. Instability means that the oil rings skip, skew, misalign, and abrade. It would be a highly desirable maintenance practice to measure a new ring at assembly and again measure it at the time of repair. The width difference represents lost metal; the lost metal will become an oil contaminant and cause the bearings to fail prematurely.

To get oil rings to function as designed, the shaft system must be nearly perfectly horizontal. Ring immersion in the lubricant must be in the right range—usually close to 5/32 inch or 8–10 mm below the oil level. Moreover, to avoid ring abrasion and dangerous oil contamination, ring eccentricity must be within 0.002 inch (0.05 mm), and surface finish should be reasonably close to 32 and, at most, 64 root mean squared (RMS). Oil viscosity should be close to typical ISO VG 32, and temperatures must be in the moderate range. These different and equally important parameters are rarely all within their respective desirable ranges in actual operating plants. If several of the individual parameters are just "borderline acceptable," oil rings will intermittently malfunction.

Slinger Disk

Because the behavior of oil rings is very difficult to control, some reliability-focused purchasers try to avoid them. As oil rings move downhill, they will inevitably contact either the edge of a shaft groove or the inside of a bearing housing. Such contact not only will tend to slow down oil rings but also will very often cause the oil rings to undergo abrasive wear. In such a case, slivers of slinger ring material will contaminate the lubricant (generally making the oil appear gray) and cause premature bearing failure. This explains why flinger disks should be used whenever possible, although they, too, will have to be used within manufacturer-approved peripheral speeds. Flexible flinger disks can be trimmed to the required diameter, and by virtue of the failure risks associated with them, flinger disks are highly cost effective.

Slinger disks are attached directly to the shaft and are designed to pick up the oil and splash it throughout the bearing housing. This can be very effective in maintaining lubricant temperature and quantity supplied to the bearing.

FACTORS THAT AFFECT OIL SUMP LUBRICATION

The most critical elements of lubrication are quality and quantity. Without one, the other is negatively affected. Having the proper quantity of poor-quality oil is no better than having an insufficient quantity of high-quality oil. Selection of the best oil for the application also should be considered an important factor in optimizing equipment lubrication.

Quantity

Having the proper quantity of oil is possibly even more important than maintaining the quality of the oil. Oil sump lubrication does not require that a specific level be maintained for proper "loading" of the bearing. However, if the level of oil in the sump reaches critically low or high points, damaging conditions may occur. In a low-level operating condition, the bearing will not receive enough lubricant for proper film strength—a precursor of surface contact, skidding, and possibly catastrophic failure. In a high-level oil operating condition, churning of the lubricant will occur, accelerating the oxidation rate as a result of excessive air and elevated temperatures.

Oil Starvation

Too little lubricant can be catastrophic. This is commonly the result of incorrect filling, oiler settings, and unrecognized leakage. Without enough oil to prevent friction, *thermal runaway* can happen very quickly to a steel bearing. As the temperature of the bearing increases, the ball and race both expand, which creates an even tighter fit. This increases the temperature even more, and this cycle continues to a rapid, catastrophic failure. A less obvious cause of oil starvation is high viscosity—as a result of oxidation or degradation or improper oil selection. If the oil is too thick, it cannot penetrate the small clearances of a rolling-element bearing, particularly at higher speeds.

Excessive Lubrication

It is a common mistake to believe that more is better—especially when it comes to oil sump lubrication. Too much oil can affect the operation of flinger rings, slingers, and direct bearing contact. Churning can lead to higher operating temperature, increased oxidation, and reduced equipment efficiency. Another result of high lubricant levels is leaking seals. Many bearing housing seals are designed for use with proper lubricant levels.

In these arrangements, the normal oil level is set at around one-third to one-half the diameter of the rolling-element ball (or roller), as shown in Figure 4.4.

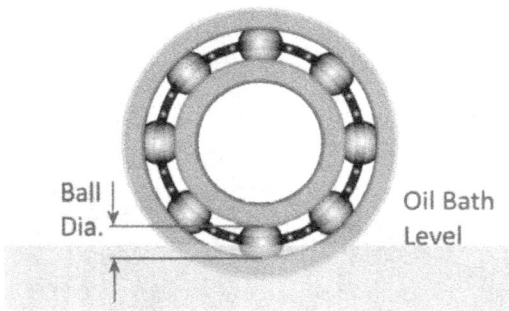

Ball Dia.

Oil Bath Level

FIGURE 4.4 Oil bath lubrication showing a typical oil level.

Quality

In basic terms, quality of lubrication can be looked at two ways:

1. How the lubricant can become contaminated, and
2. How the lubricant can degrade.

Although contamination is widely recognized for its effect on the quality of oil, degradation can be just as damaging to equipment. The leading causes of contamination are particulate matter, moisture, incompatible fluids, and air entrainment. The leading causes of degradation are oxidation, heat, and use.

POOR OIL SUMP LUBRICATION—CAUSES

Water Contamination

Water is one of the main culprits that shorten bearing life or contaminate oil. Also, thermal cycling caused by temperature changes tends to draw airborne contaminants into the bearing housing. Damaged or worn seals allow product to enter the housing and create severe problems. Water contamination of oil can cause several problems relative to oil contamination or degradation. Water in oil will dramatically decrease the effectiveness of the lubricant and therefore reduce service life. Typical recommendations, depending on the source, are to maintain water at 100 or 200 parts per million (ppm) or less in the lubricating oil. This can be a challenge in some applications but should still be a target. Whereas there is no widely recognized way of calculating the effect of moisture on bearing service life, testing has shown a dramatic drop in life as water content increases above that target.

For oil bath applications, verifying moisture content should be part of periodic oil testing. If the moisture content is too high, the oil should be replaced. In circulating oil applications, dewatering can be part of the circulating oil system so that the water content of the oil can be maintained at a suitably low level. One of the major routes for water to gain access to the bearing housing is intake of moist air into the housing when the pump is shut down and cools. This moist air cools as the pump cools, potentially releasing condensation into the bearing housing.

Because each type of oil has its own safe level of water before damage occurs, the common practice of measuring parts per million is not conclusive. There are significant differences between oils, beginning with mineral oil and synthetic bases. Additive packages can also make a difference in how much water an oil can hold before phase separation occurs—and free water forms.

Temperature also plays a major role in how much water an oil can hold. Damaging levels of water, or free water, begin to occur in some mineral oils at between 400 and 500 ppm at 60°C (140°F). Free water may form at 200 ppm at 51.7°C (125°F) in the same oil. Therefore, it is important to know the saturation point of an oil at a given temperature to begin to determine a valuable set point for effective lubrication maintenance. By the time water becomes visible, damage is already occurring to both the oil and the surfaces of the equipment and components (Figure 4.5).

Parts per Million	Effect	Detection
1300		
1000	Extremely damaging to steel element bearings	Visibly cloudy oil
600	Damaging to bearings	Being sensed at damaging levels. Karl Fischer and centrifugal detection are typically done in laboratories.
200	Very safe, non-damaging	Water sensing detects levels online
100	New oil levels	

FIGURE 4.5   Moisture in oil detection.

Particle Contamination

Even though bearings are made of hardened steel, they are surprisingly susceptible to contamination. The reason is the very high contact pressures that are present in the small contact zones between the rolling elements and the raceway. Pressures of around 200,000 lb/in^2 between the ball and the race in an angular contact ball bearing are not uncommon. Overrolling contaminant particles with such high contact pressures result in dents in the raceway. Particulate contaminants, whether those particles are soft or hard, large or small, will cause some damage to bearings.

The presence of contamination shortens bearing service life in two ways. Every time a rolling element passes over a dent, contact pressure increases at the edge of the dent. Higher stresses result in shorter fatigue life. The second mechanism is wear. Although the balls do roll in a ball bearing, because of the curvature of the balls and races, some sliding occurs as well. The sliding portions of the contact, when contamination particles are present, can result in wear of the surfaces.

Bearing housings "breathe" because rising temperatures during operation cause gas volume expansion and dropping temperatures at night or after shutdown cause gas volume contraction. This thermal cycling caused by temperature changes tends to draw airborne contaminants into the bearing housing. Damaged or worn seals allow moisture-laden contaminated air to enter the housing and create severe problems.

Airflow across pumps generated by fans, blowers, and even the pump motor can be sufficient to create a pressure differential between the bearing housing and the surrounding atmosphere. This will increase the intake rate into the bearing housing as much as 10 times, leading to higher levels of contamination. Steam quenching and high-pressure washing can significantly increase the rate of intake of moisture into bearing housings. Care should be taken to avoid direct spray or steam around vents, oilers, and seals. Many vents, oilers, and seals are designed to prevent this intake—but they do need to be specified.

Operating oil rings on rotating shaft systems that are not horizontal will cause the bronze slinger ring to spin and rub against the low side of the housing, resulting in severe wear on the ring. The resulting bronze particles can clearly damage the bearings.

Particle contamination is probably the best known form of lubricant contamination. This form is considered the cause of wear of component parts and surface fatigue. In a study done by the National Research Council of Canada on the effects of particle contamination, nearly 85 percent of contaminant wear of components and surfaces was induced by particles. To make matters worse, particle contamination can create more particles—and thus more wear.

Lower particle counts significantly extend the life expectancy of equipment. For example, by reducing contamination levels from ISO 21/18 to ISO 14/11, life of a 50 gal/min pump could be extended by a factor of 7. Particle contamination can occur as a result of intake from the surroundings, improper cleaning of the bearing housing during maintenance cycles, and corrosion products from the high water content in the oil.

Heat

Elevated operating temperatures are a major contributor to oil oxidation. Combined with air, particulate, and water contamination, the chain reaction of oil oxidation begins. Additives are affected first, followed by the base stock, which leads to machine and component surface wear and fatigue. For every 7.8°C (18°F) increase in oil operating temperature, the oxidation rate doubles. Oil operating at 75°C (167°F) will last 100 times longer oil operating than at 130°C (266°F) (Figure 4.6).

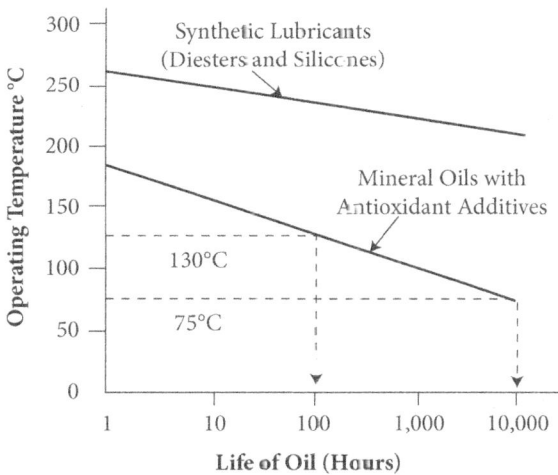

FIGURE 4.6   Oil life varies with base oil type.

Air Entrainment

Air entrainment is the primary source of oxygen in the oxidation failure of oil. New oil can contain as much as 10 percent air at atmospheric pressure. Splash-lubricated gearboxes, bearing housings using flinger rings or slingers, and compressors are all aeration-prone applications. Excessive aeration has a negative effect on acid number, oil color, film strength, and viscosity. In addition, air entrainment can lead to accelerated surface corrosion, higher operating temperatures, and oil varnishing.

Oil Degradation

The primary causes of oil degradation are high heat, air entrainment, and mixing incompatible fluids. Increased viscosity (thickening) is one of the results of this degradation. This usually happens over time and varies by the combination of these elements (Figure 4.7).

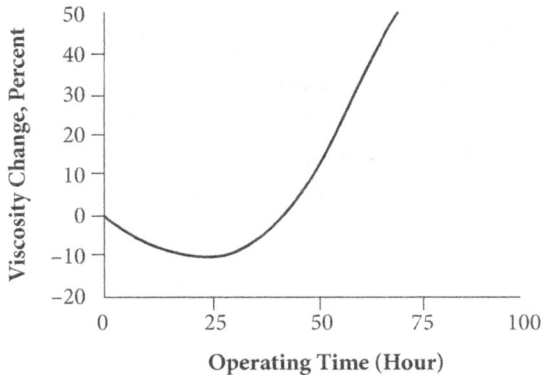

FIGURE 4.7   Viscosity effect as indicator.

PREVENTION MEASURES TO IMPROVE LUBRICATION

Preventing Improper Quantity

Maintaining the proper quantity of a lubricant is perhaps the easiest means of increasing lubrication effectiveness and life. Consult with your equipment manufacturer or the operating manual for recommended oil levels, optimal lubricating equipment, and preferred practices. Most equipment will have an external marking indicating proper oil level that is either cast into the housing or marked on an attached tag. In a low-level operating condition, the bearing will not receive enough lubricant for proper film strength—leading to catastrophic failure. As the temperature of the bearing increases, the ball and race both expand, which creates an even tighter fit. This increases the temperature even more, and the cycle continues to a rapid, catastrophic failure.

In a high-level operating environment, churning of the lubricant will occur, accelerating the oxidation rate as a result of excessive air and elevated temperatures. It is a common mistake to believe that more is better—especially when it comes to oil sump lubrication. Too much oil can affect the operation of oil rings and flingers. Another result of high lubricant levels is leaking seals.

One of the most widely used methods of maintaining the proper level lubricant in a bearing housing is the constant-level oiler (see Figure 4.2). This type of oiler replenishes oil lost by leakage through seals, vents, and various connections and plugs into the bearing housing. Once the proper level has been set, replacing the oil in the reservoir is the only required maintenance.

For smaller bearing arrangements and slower speeds, oil bath lubrication arrangements are commonly employed. In rolling-element bearings where oil bath lubrication is used, the normal oil level is set at around one-third to one-half the diameter of the rolling-element ball (or roller), as shown in Figure 4.4. But where flinger rings are used, the normal oil level depends on the shaft speed relative to the depth of submersion. Hence, a good rule of thumb to use is to allow full thickness of the ring (e.g., $3/16$ inch) at the deepest point. Slinger disks are less susceptible to problems of overlubrication because they are attached directly to the rotating shaft. However, they have been known to cause misfeeding of constant-level oilers through the creation of hydraulic currents in the oil sump.

Preventing Contamination

Housing components including oilers, seals, and vents, when specified properly, can be very effective in preventing contamination (Figure 4.8). Constant-level oilers, which are used in maintaining oil levels, are vented to the surrounding atmosphere and can introduce contamination into the sump housing. By switching to a nonvented oiler, intake can be significantly reduced. Bearing housing seals, more often recognized as *isolators*, have been producing positive results in reducing oil leakage and intake of contamination. Labyrinth-type isolators are used most widely on horizontal pumps and are designed to prevent contamination. Designed to allow increased pressure created by normal pump operation to vent through the seal, labyrinth (laby) seals have proven to be very effective at reducing and sometimes eliminating contamination. Lip seals also can be very good at preventing contamination, but a contacting-type design requires more frequent replacement to ensure close-tolerance operation.

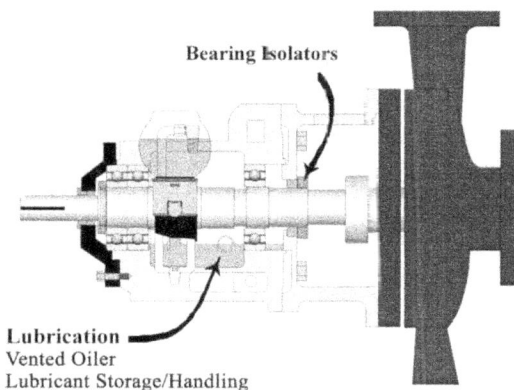

FIGURE 4.8   Contamination sources.

Prevention of Degradation

The life of a lubricant is significantly reduced when the lubricant is exposed to high operating temperatures. As mentioned previously the oxidation rate of oil doubles every 7.8°C (18°F). This can be significant when considering that pump operating temperatures are frequently near or above 60°C (140°F). By simply lowering the operating temperature of the oil to 50°C (22°F), a 50 percent reduction in the rate of oxidation can be realized—doubling the effective life of the oil. The basic methods to reduce (or maintain) lower oil operating temperatures include:

- **Use the correct oil viscosity.** Too high or too low will raise the temperature of the oil.
- **Use quality oil.** Do not buy cheap oil to save money—it will end up costing you more.
- **Use the right amount of oil.** Maintain proper oil levels—too much or not enough will increase the temperature of the oil.
- **Keep the oil clean.** Contaminated oil operates at a higher temperature than clean oil.

Preventing Operating Temperature Problems

The recommended oil operating temperature range for a particular application is usually specified by the equipment manufacturer. Exceeding the recommended range may reduce the oil's viscosity, resulting in inadequate lubrication. Subjecting oil to high temperatures also increases the oxidation rate. As noted earlier, for every 10°C (18°F) above 66°C (150°F), an oil's oxidation rate doubles, and the oil's life is essentially cut in half. Longevity is especially critical for turbines in hydroelectric generating units, where the oil life expectancy is several years. Ideally, the oil should operate between 50 and 60°C (120 and 140°F). Consistent operation above this range may indicate a problem such as misalignment or tight bearings. Adverse conditions of this nature should be verified and corrected. Furthermore, when operating at higher temperatures, the oil's neutralization (acid) number should be checked more frequently than dictated by normal operating temperatures. An increase in the neutralization number indicates that the oxidation inhibitors have been consumed and the oil is beginning to oxidize. The lubricant manufacturer should be contacted for recommendations on the continued use of the oil when the operating temperatures for a specific lubricant are unknown.

HOW TO STOP THE CONTAMINATION

Unless a pump is provided with suitable bearing housing seals, an interchange of internal and external air (called *breathing*) takes place during alternating periods of operation and shutdown. Bearing housings breathe because rising temperatures during operation cause gas volume expansion, and dropping temperatures at night or after shutdown cause gas volume contraction. To stop this breathing and resulting contamination, there should be no interchange between the interior air and the surrounding ambient air of the housing. Open or inadequately sealed bearing housings and non-pressure-balanced oilers promote this back-and-forth movement of moisture-laden, contaminated air.

Sealing of Bearing Housings

Sealing of bearing housings is critical to prevent the intake of contaminants. Seals should be appropriate for the operating conditions and should be designed to prevent entry of all types of contaminants into the bearing system. Three types of seals that may be used are lip-type contact seals, labyrinth seals, and magnetically charged face seals. Lip seals will seal only while the elastomer material (the lip) makes full sliding contact with the shaft (see Figure 4.9). However, the seal lip contact with the shaft must be lubricated, and even with lubrication, the seals do create added frictional heat. Over a period of time, lip seals will wear and, depending on conditions, may fail well before a bearing fails. Seal failure will likely lead to bearing failure.

Operating at typical shaft speeds on process pumps, lip seals show leakage after about 2,000 operating hours. To prevent contaminant intake, lip seals should be replaced just before they fail—perhaps four times a year to be safe. In sharp contrast, modern rotating labyrinth seals run for longer periods of time.

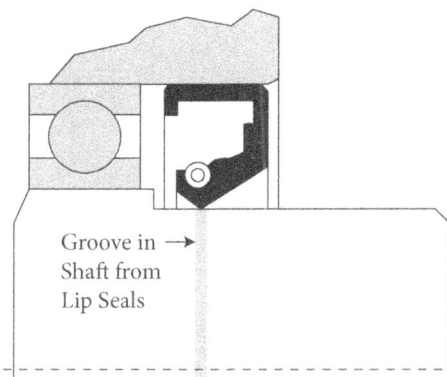

FIGURE 4.9 Lip seal.

Labyrinth seals may come in a variety of forms but certainly one of the more common and effective is the cartridge-type labyrinth seal (often referred to as an *isolator seal*) that includes a rotating element that is mounted on the shaft and a stationary element in the housing (see Figure 4.10). The rotor provides a flinger effect, throwing contaminants off the seal when it is rotating. The axial or face labyrinth created by the nesting of the rotor and stator makes it extremely difficult for any type of contamination to pass through the seal. Additionally, there are typically ports at the bottom of the seal to allow any contamination that has entered the labyrinth to exit through these ports.

Magnetically charged face seals also may be a viable option to seal bearing housings. These seals use magnets to provide the force necessary to keep precision-lapped face seal surfaces in contact. Similar to the isolator-type seal, one of these face seal surfaces is part of a rotor driven by the shaft and the other is part of a stationary body attached to the housing. The magnetic force compresses and aligns the mating surfaces so that the seal is perfectly adjusted for life. These seals can prevent intake of contaminant (including water wash-down) as well as retain lubricant. Although there is contact between the face seal surfaces, seal life is still substantially longer than that of lip seals. Some friction is created, but heat generation should be less than that of a lip seal but more than that of an isolator seal.

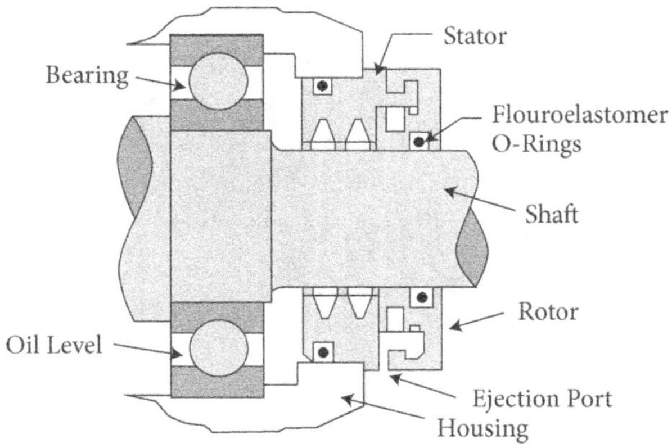

FIGURE 4.10   Isolator-type labyrinth seal.

Oilers

In order to combat the potential for oil contamination, the closed-system oiler was developed in place of the constant-level oiler. Constant-level oilers are designed to maintain a predetermined oil level in a sump, which is necessary for proper lubrication. If the oil level were to drop below the threshold point, the depleted oil would automatically be replenished by the lubricator, returning it to its original level (Figure 4.11).

In order to combat the potential for oil contamination, the closed-system oiler was developed. This type of oiler is effective in minimizing and eliminating the intake of contaminants into the oil sump, especially in dirty environments. Some closed-system oilers contain a pressure balancing line that is connected from the headspace of the oil sump to an air chamber built into the surge body of the oiler. This air chamber is sealed from the outside atmosphere in order to prevent the intake of contaminants. Additional types of closed-system oilers are available that mount directly on the centerline of the oil level to be maintained.

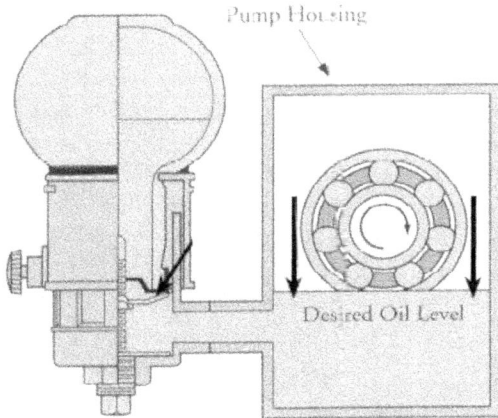

FIGURE 4.11   Constant-level (non-pressure-balanced) lubricator.

Pressure-balanced oilers (Figure 4.12) decrease downtime risk. They differ from the nonbalanced type by incorporating an external pressure balance pipe so as to make sure that the pressure inside the bearing housing and the pressure at the tip of the wing nut in the constant-level lubricator are always identical. Consequently, the oil in the bearing housing is pushed downward by the hot gas (air) with the *same* pressure that is pushing downward on the oil in the oiler, and thus there is no change in the oil level.

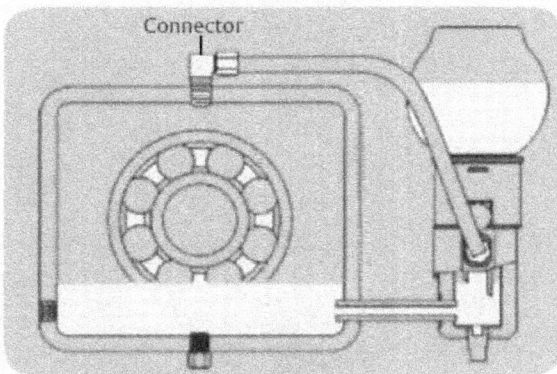

FIGURE 4.12   Pressure-balanced constant-level lubricator.

With the use of constant-level oilers, maintenance efficiencies can be increased while minimizing maintenance costs and the loss of production time. Most constant-level oilers are adjustable, allowing for use in many applications. However, there are oilers that do not allow for fluid level adjustability, eliminating potential installation errors.

CONCLUSION

Pumps are an integral part of almost every plant. The key to their reliability depends on the lubrication system. Operators and lubrication technicians are vital players in this journey. Best practices for operators in early lubrication-related mechanical-problem resolution include monitoring of oil levels, color, foaming, and cleanliness; noting changes in vibration, unusual sounds, or oil leaks; and vigilant contamination-control practices. These good practices can enhance pump reliability and extend pump life.

ICML Questions

1. The most important criterion in the selection of a proper pump lubricant is
   a. viscosity.
   b. additives.
   c. foaming property.
   d. All of the above

2. Too much of oil in the oil sump
   a. shortens the effective life of the oil.
   b. cools the bearing.
   c. reduces the oxidation property of the oil.
   d. All of the above

3. The most important consideration in oil sump lubrication is
   a. speed.
   b. oil viscosity.
   c. load.
   d. All of the above

4. The normal oil level in oil sump lubrication is set at around
   a. one-third to one-half the diameter of the rolling-element ball or roller.
   b. one-half to two-thirds the diameter of the rolling-element ball or roller.
   c. one-fourth to one-third the diameter of the rolling-element ball or roller.
   d. Any amount is okay.

5. Damaging levels of water, or free water, begin to occur in some mineral oils at 60°C at
   a. between 400 and 500 ppm.
   b. between 100 and 200 ppm.
   c. between 50 and 100 ppm.
   d. between 0 and 50 ppm.

6. The primary cause of oil degradation is
   a. high heat.
   b. air entrainment.
   c. mixing of incompatible fluids.
   d. All of the above

7. The basic method to reduce (or maintain) lower oil operating temperatures is to
   a. use the correct oil viscosity.
   b. use the right amount of oil.
   c. keep the oil clean.
   d. All of the above

CHAPTER 5

Lubrication of Reciprocating Compressors

A *reciprocating compressor* is a positive-displacement machine that uses pistons driven by a crankshaft to deliver gases at high pressure. Reciprocating compressors are often the most critical and expensive systems at a production facility, and thus they deserve special attention. Applications include oil refineries, gas pipelines, chemical plants, natural gas processing plants, and refrigeration plants.

The main parts of a reciprocating compressor are a cylinder, a piston, piston rings, inlet valve, discharge valve, and a drive assembly consisting of a crankshaft and the connecting rod. *Compression* in reciprocating compressors occurs within the cylinder as a four-part cycle such as intake, compression, expansion, and discharge. The cycle is repeated as the piston moves back and forth. For each cycle, gas is drawn into the cylinder, compressed, and delivered to the discharge piping. The piston rings maintain a seal between the piston and the cylinder, which lets the gas be compressed without leaking past the piston. In a reciprocating compressor, compressing a gas causes its temperature to rise. When more gas is compressed, the higher will be the final temperature. When high discharge pressures are required, compression is carried out in two or more stages, and as a result, the outlet gas temperature increases.

To overcome this temperature increase, the gas is cooled between stages to limit the temperature to acceptable levels. Interstage cooling help improves compressor efficiency and reduces power consumption for the range of operating temperatures. In this chapter, lubrication of reciprocating compressors in terms of oil type, oil viscosity, major lubrication areas, and force-feed lubricators are described in subsequent chapters.

Lubrication is vital for successful operation of a compressor. Adequate lubrication of the proper type in the proper place at the proper time is the key to successful lubrication. Although lubrication is an ongoing concern throughout the life of the compressor, the initial package design is very important and deserves special attention.

There are two independent systems for lubricating compressors: the *frame oil system* and the *force-feed system*. The frame oil system is a pressurized circulating system that supplies oil under constant pressure to the crankshaft, connecting rods, and crossheads. The force-feed system is a low-volume, high-pressure injection system that supplies small quantities of oil at regular intervals to lubricate the piston rod packings and piston rings. In many applications, these two systems can use the same lubricant, but in some applications, the lubricants must be different. In both bearings and cylinders, the lubricating oil must form and maintain a strong film that will minimize friction and wear

BEARING LUBRICATION

In reciprocating compressors, the oil sump for lubrication of bearings is contained in a reservoir in the base of the crankcase. A number of methods are employed for delivering oil from the reservoir to the lubricated parts.

Splash Lubrication

Oil may be delivered to lubricated parts by splash lubrication. In this type of lubrication, a projection from one or more cranks or connecting rods dips into the oil and produces a spray that reaches all internal parts. In splash lubrication, the level of oil in the reservoir should be maintained within predetermined limits. Over- and underlubrication may cause problems.

Many horizontal compressors have a flood system for bearing and crosshead lubrication. Oil is lifted from the reservoir by disks on the crankshaft and is removed by scrapers. The oil is then directed to the bearings by passages or is allowed to cascade down over the crosshead bearing surfaces.

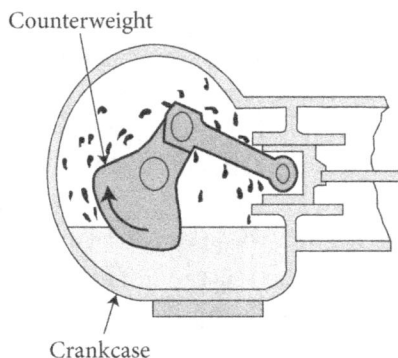

FIGURE 5.1   Oil is lifted from the reservoir by the disks of the crankshaft.

Pressure Circulation Lubrication

The frame oil system uses an oil pump that is attached to the compressor and driven from the compressor shaft. This pump is normally a gear pump or a screw pump. It sucks oil from the reservoir (oil sump) and delivers it through a filter usually 25 μm to an oil cooler. From the cooler, the oil is distributed to the top of each main bearing and crosshead and to the outboard bearings. The oil fed to the main bearings is picked up at the main bearing journals and carried to the crank journals through the drilled holes in the crankshaft. Lubricating oil from the crank bearings up to the piston pin bushings travel through the hole drilled in the connecting rods. As oil is forced out from various bearings, it is collected back into the oil sump, which is located in the base of the compressor. The oil is recirculated, and the process is repeated. Oil from the outboard bearings is carried back to the sump by drain lines. Oil scraper rings in the frame end prevent oil leakage out along the piston rod. Because of this difficult passage of oil in the drilled holes in the connecting rod, prelubrication is required before startup because it takes time to fully lubricate the components. This is accomplished with an auxiliary lube pump.

Nonlubricated reciprocating compressors have lubricated running gear (shaft and bearings) but no lubrication for the cylinder, packing, and valves. This design produces oil-free air. In this system, a positive-displacement pump draws oil from the reservoir and delivers it under pressure to the main bearings and the crankpin bearings and hence to the crosshead pin bearings and crossheads. Lubrication in this case is accomplished either by a pump driven from the crank end or by a separately mounted pump.

Pressure circulation systems are equipped with suction strainers and pressure-relief or control valves. Oil filters, usually of the full-flow type, and pressure gauges are also provided. Crankcase oil heaters are specified for outdoor compressors to keep the oil at required viscosity and to prevent condensation with resulting corrosion. Therefore, when using crankcase heaters on a compressor that is not in operation, the auxiliary lube pump should be run continuously; otherwise, local overheating and carbonization of oil will occur.

FRAME LUBRICATION SYSTEM

Components

A complete frame lubrication system is shown in Figure 5.2.

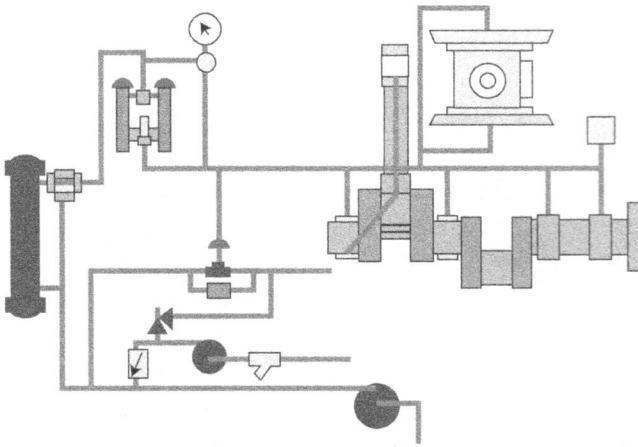

FIGURE 5.2   Frame lubrication system.

Oil Strainer

An oil strainer, installed upstream of the pump, will prevent very large particles and objects from getting into the pump. Normally, this is a 30–40 mesh strainer.

Oil Pump

The oil pump constantly supplies oil to all the journal bearings, bushings, and crosshead sliding surfaces. The pump is direct coupled to the crankshaft by a chain and sprocket and is designed to provide adequate oil flow to the bearings when the compressor is operating at one-half maximum rated speed. The compressor frame-driven lube oil pump maintain oil pressure with a spring-loaded regulating valve within the pump head. Lube system pressure can be raised or lowered by adjusting the spring tension on this valve.

Oil Cooler

An oil cooler is required to maintain the oil temperature within acceptable limits. Factors that must be taken into account while sizing a cooler are the cooling medium, temperature of the cooling medium, cooling medium flow rate, lube oil temperature, lube oil flow rate, and oil heat rejection data. An insufficient cooler water flow rate is the primary cause of high oil temperatures. The cooler should be mounted as close to the compressor as possible, with piping of adequate size to minimize pressure drop of both the lubricating oil and the cooling medium.

Oil Temperature Control Valve

Thermostatic valves are required in conjunction with the cooler to control the oil temperature in the compressor. A thermostatic valve is a three-way valve that has a temperature-sensitive element. As the oil is heated, the sensing element will open the third port in the valve (Figure 5.3).

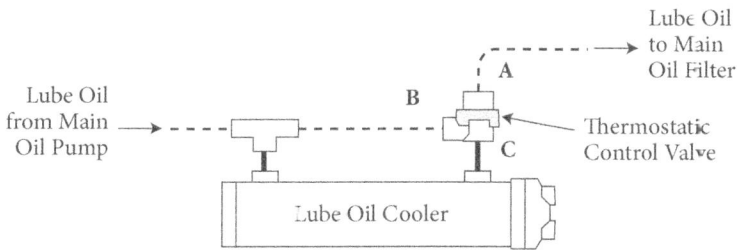

FIGURE 5.3 Thermostatic valve in mixing mode. The thermostatic control valve configuration may vary from this schematic depending on valve size. In mixing mode B, the connection is lube oil from the main oil pump with a tee connection to the lube oil cooler inlet; C is from the lube oil cooler outlet; and A is to the main oil filter. Valve connections A, B, and C are marked on the valve.

There are two configurations for a thermostatic valve: a diverting mode and a mixing mode. For a thermostatic valve in mixing mode, as the oil heats up, the element opens a port in the thermostatic valve that allows oil from the cooler to mix with the hot oil from the bypass. In the diverting mode, the oil is diverted to the cooler when the oil from the compressor is hot enough to open the valve. The diverting mode monitors the temperature of the oil coming out of the compressor. The mixing mode monitors the temperature of the oil going into the compressor.

Oil Filters

Oil filters are required on the compressor frames to remove contamination that can damage the equipment and contamination that can damage the oil. Contaminants that damage the equipment include:

- Wear particles from the equipment
- Airborne particles such as dust and sand
- Solid particles from the gas stream
- Dirt from new oil

Contaminants that damage the oil include:

- Oxidized oil components
- Air bubbles
- Moisture from air
- High oil temperature

The compressors are equipped with simplex cartridge style pleated synthetic filters as standard

Compressor Prelube Pump/Auxiliary Oil Pump

An automated compressor prelube system is recommended to provide oil flow prior to startup, which will extend bearing life and avoid damage to the compressor. The purpose of the prelube pump is to be sure that there is oil flow to all bearings and bushings and that the clearances are filled with oil prior to startup. A start permissive is desirable to sense the minimum required pressure at the inlet to the oil header. Sometimes an auxiliary oil pump acts as the prelube pump in the compressor.

Oil Heaters

Frame oil heaters may be needed if the compressor must be started in cold weather. Depending on the operational environment of the compressor, one possible heating mode is to simply start the compressor immediately if needed; another mode is to heat the oil from ambient to a minimum temperature prior to starting. The specific application requirements will determine which heating mode is necessary.

Cold Starting

If a compressor is exposed to cold ambient temperatures, the oil system must be designed so that the unit may be safely started with adequate oil flow to the journal bearings. Temperature-controlled cooler bypass valves, oil heaters, and cooler louvers may be needed to ensure successful operation.

Lube Oil Analysis Sampling Point

A sampling point should be installed at a convenient location between the oil pump and filter. This should be installed at an easily accessible location and designed to minimize the amount of dirt or debris that can collect around the dispensing point. A needle valve should be used to better control the flow of pressurized oil.

Frame Lubrication Operating Conditions

Compressor Frame Lubricant

Normally, it is recommended to use an ISO 150 grade paraffinic mineral oil that provides oxidation inhibition, rust and corrosion inhibition, and antiwear properties in the compressor frame. This is commonly called *rust- and oxidation-inhibited (R&O) oil*.

In limited circumstances and with prior approval, cold weather installations may use multiviscosity oils in the compressor frame if the oil supplier can certify that the oil is shear stable. The viscosity of shear-stable oil degrades less with use. Most oil suppliers will certify their oil as shear stable if the viscosity degrades less than a certain percentage in specific tests.

As a result, multiviscosity oils are subject to a shorter oil life than single straight grade oils by 30 to 50 percent.

Synthetics such as polyalkylene glycol (PAG), polyalphaolefin (PAO), and ester-based lubricants are acceptable in the compressor frame, provided that they meet the operating viscosity requirements as outlined next. Compounded lubricants are prohibited in compressor frames.

Synthetic Compressor Lubricants.   Synthetic oils have become more widely used and accepted as compressor lubricants because of their higher *autoignition* characteristics and ability to prevent carbon buildup on valves and piston rings. The use of synthetic oils may allow the reduction of feed rates to the cylinder by approximately one-third. This reduction in feed rate will result in less oil in the downstream piping system. Reduced oil accumulation and the fire-resistant characteristics of synthetic oils help to prevent fires in the discharge lines. However, no oil is fireproof or explosion-proof.

Oil Viscosity
The minimum allowable viscosity of the oil going into the compressor should be based on the manufacturer's recommendation depending on service and environmental condition.

Oil Temperature
A minimum lube oil operating temperature (normally 66°C) must be maintained. This is the minimum temperature required to drive off water vapor. Maximum allowable oil temperature in the compressor frame is normally 85–90°C (190°F). A thermostatic control valve is set to 77°C (170°F). The oil should be maintained as close to this temperature as possible. Higher temperatures increase the oxidation rate of the oil. For every 10°C (18°F) over 66°C (150°F), the oxidation rate of the oil doubles.

Oil Maintenance
Compressor frame lubricating oil should be changed as indicated in the regular maintenance intervals or with a filter change or when oil analysis results indicate the need. A more frequent oil change interval may be required if the compressor is operating in an extremely dirty environment without sampling and analysis or if the oil supplier recommends it.

Oil Sampling
Oil samples should be collected on a regular basis and analyzed to verify the suitability of the oil for continued service. Consistent oil analysis can identify when to change the oil on the basis of need rather than on a scheduled interval. Depending on the service, the length of time between oil changes can be significantly extended via oil analysis. Oil analysis should include:

- **Viscosity testing.** This should be performed at 40°C (100°F) and 100°C (212°F). This is to be sure that the oil has not mixed with cylinder oils or process gas.
- **Particle counting.** Based on the latest version of ISO 4406.
- **Spectroscopy.** To determine where metals, contaminants, and additives are located.
- **Fourier-transform infrared spectroscopy (FTIR).** To check for oxidation, water or coolant contamination, and additive depletion. This is more important if a separate lube oil is used for the force-feed system.

Oil System Cleanliness

The compressor frame oil system and components must be free of foreign matter, including, but not limited to, dirt, sand, rust, mill scale, metal chips, weld spatter, grease, and paint. It is also recommended that after major overhauling, all oil piping systems must be flushed using pump and filtered, clean production oil. All compressor frame cavities must be thoroughly cleaned prior to assembly, and compressors must be test run with a filtered loop lube system at the factory.

Prior to assembling the lube oil piping, remove scale, weld slag, rust, and any other matter that could contaminate the lube oil:

- The entire lube oil system must be complete and closed.
- The crankcase should be filled to the correct level with the appropriate oil.
- The proper lube oil filters must be installed correctly.
- The oil pressure transducer or gauge, oil filter differential pressure transducers or gauges, and the oil temperature resistance temperature detector (RTD) or indicator must be operational and the values viewable.

Start the prelube pump, and record the oil pressure, oil filter differential pressure, and oil temperature.

A continuous hour of prelube flushing time must be achieved with an oil filter differential pressure increase of less than 10 percent of the measured oil pressure into the filter. Record the oil pressure, oil filter differential pressure, and oil temperature at 15-minute intervals. If the oil temperature increase is greater than 10°C (18°F) during an hour of flush time, it is not a valid test of system cleanliness owing to oil viscosity change.

If the differential pressure or temperature increase exceeds the limits, after an hour of prelube flushing, continue the flushing operation. Wherever the lube oil filter differential pressure exceeds the change filter limits, stop the prelube pump and change the oil filter. Reset the time, and continue the flushing operation until a continuous hour of flushing time is achieved within the differential pressure and temperature increase limits to ensure system cleanliness.

It is recommended that the lube oil piping downstream of the installed oil filter not be disturbed because contaminants that enter that piping or open ports will be flushed into the bearings, causing catastrophic damage. If the piping must be removed or altered, great care must be taken to cover the inlets to the oil header, the ends of the piping, and the filter outlet so that no contaminants may enter. Before reinstallation, chemical and mechanical cleaning is required. The pipe must then be flushed in accordance with stated cleanliness requirements.

For all compressors with an oil piping system greater than 50 ft (15 m), cleaning and flushing must result in a cleanliness level to ISO 4406, Grade 13/10/9, and National Aerospace standard (NAS) 1638, Class 5, prior to starting the compressor.

See Table 5.1 and ISO 4406 and/or NAS 1638 for cleanliness requirements for parts used in hydraulic systems for complete information. Use a competent oil laboratory for sample testing.

TABLE 5.1   Oil Flushing Cleanliness Requirements

ISO 4406, Grade 13/10/9		
Grade Requirements	Particle Size, pm/ml of Oil Sample	Number of Particles Allowed
/13	>4	40–80
/10	>6	5–10
/9	>14	2.5–5
NAS 1638, Grade 5		
Particle Size Range, pm/100 ml Oil Sample	Grade 5 Maximum Number of Particles	
5–15	8,000	
15–25	1,424	
25–50	253	
50–100	45	
>100	8	

Selecting Lubricants

The major factors involved in the selection of reciprocating compressor lubricants include:

- Type, size, and speed of compressor
- Gas being compressed
- Number of stages
- Pressure and temperature at each stage

▪ Environment
▪ Type of lubrication system

Most reciprocating compressors are of the one- or two-stage type, with smaller numbers of multistage machines—three, four, or more stages. From a lubrication point of view, one- and two-stage machines generally are similar, whereas multistage units may have somewhat different requirements depending on pressures, temperatures, gas conditions, and the size and speed of the pistons.

Factors That Affect Crankcase Lubrication

The factors that affect reciprocating compressor bearing lubrication are load, speed, temperature, and the presence of water and other contaminants. The most important requirement for crankcase lubrication is that the oil is of suitable viscosity at operating temperature. At normal ambient temperatures, the oil should have a minimum viscosity of 200 Saybolt universal seconds (SUS) at 130°F (266°F). This is equivalent to a Society of Automobile Engineers (SAE) Grade 30 or ISO Grade 100 lubricant. The type of oil used for the lubrication of bearings and running-gear components in reciprocating compressors must comply with the compressor manufacturer's recommendations. The oil in circulation within the compressor crankcases is broken up into a fine spray or mist by splash or oil thrown from the rotating parts. Hence the oil should display good antifoaming qualities. The antifoaming properties are particularly important in compressors using splash or flood systems. It should be of good-quality nondetergent mineral oil that should contain rust and oxidation inhibitors.

Problems with Crankcase Oil

Oxidation.   As oil is broken up into fine spray or mist by splash, a large surface of oil is exposed to the oxidizing influence of warm air. Hence oxidation takes place at a rate that is a function of the operating temperature and the ability of the oil to resist chemical change. Oil oxidation is accompanied by a gradual increase in viscosity and, eventually, by the deposition of insoluble products in the form of gum or sludge. These deposits may accumulate in oil passages and restrict the flow of oil to bearings. Conditions that promote oxidation in crankcases are mild, however, compared with oxidizing conditions in compressor cylinders.

Water.   Although water may enter compressor crankcases by condensation from the atmosphere during idle periods or possibly from leaking jackets, there is generally little water present because of the continuous venting of water vapor at crankcase temperatures. Normally, therefore, there is little opportunity for the formation of troublesome emulsions, which could combine with dust and other contaminants to form sludge that would restrict the flow of oil to lubricated surfaces. A good compressor crankcase oil, nevertheless, will need adequate water-separating ability to resist the formation of harmful emulsions and to permit water to collect at low points where it may be drained off.

Frequency of Oil Changes

It is not practical to predefine how often an oil should be changed, The oil will become contaminated with foreign material being held in suspension as well as moisture from condensation; therefore, the time interval for oil changes is governed by the operating and local ambient conditions. It must be remembered that the oil charge will not last indefinitely, and certainly if the compressor is shut down for an overhaul, it is poor practice not to completely clean the crankcase, change filters, and install new oil at that time. Periodic oil analyses are strongly recommended in an effort to determine optimal change frequencies. In addition to moisture content and viscosity stability, total acid number (TAN) and flash point may be worth tracking and trending. Lubricating oil analysis for component wear is an excellent maintenance technique.

CYLINDER AND PACKING LUBRICATION

In contrast to the lubrication of crankcases and bearings, the lubrication system for compressor cylinders and packing must be able to reliably deliver relatively small amounts of oil at higher pressures in order to lubricate the wearing surfaces of cylinders, packing, and piston rods. For cylinder lubrication, a force-feed lubrication system that uses a positive-displacement pump that must be capable of accurately delivering, monitoring, and protecting the oil flow to each of the required lubrication points is needed. If the system fails or does not work properly, the compressor units will be seriously damaged in a short period of time during operation. These are once-through systems, and the volume of oil delivered at each point needs to be just enough for proper lubrication. Therefore, the rate of lubrication at each point is critical, and overlubrication must be avoided. Excessive oil volumes can cause fouling of valves, gumming of the packing, and accumulation in the downstream piping system.

Force-Feed Lubrication Systems

Force-feed lubrication systems provide oil to compressor cylinders and the piston rod packing. The oil is fed directly to the cylinder walls at one or more points by means of a mechanical force-feed lubricator or a centralized lubrication system. Here an individual pump feeds each point, each with an adjustable stroke and a sight glass to view the adjusted feed rate. The suction stroke of the pump pulls oil from the reservoir and discharges it down the line.

The lubricator box has its own oil reservoir to lubricate the worm gear and cam. The reservoir is self-contained and is not fed by the oil system. A sight glass on the lubricator box will show the oil level in the reservoir. There are ¼-inch tube-fitting connections in the discharge lines near the force-feed lubricator pumps through which the force-feed lubrication system may be primed. Next in the discharge lines are blow-out disks. If there is a blockage in the system, the pressure buildup will rupture the disk. Venting the system through the blow-out disk causes the no-flow shutdown switch to close.

The oil then travels to the distribution blocks. It is here that the lubricating oil is apportioned to provide the exact amounts to the cylinders and packing. The pistons in the intermediate sections of the distribution block move back and forth in a continuous cycle, forcing lubricant successively though the several outlets as long as lubricant is supplied under pressure at the inlet. Each outlet has a check valve to prevent oil from baking up in the block. And indicator on the block shows the rate at which the block is cycling. From the distribution blocks, oil travels to the cylinders and packing. The system provides 1 inch minimum (25 mm) of head at the guide and cylinder inlets to help ensure long check-valve life. Some of the oil to the packing travels through to the cylinders, but the bulk of it is drained out through the pressure-vent drain fitting on the bottom to the crosshead guide and through the atmospheric drain also in the bottom of the guide. An oil level control valve, supplied by the packager and mounted on the skid, maintains proper level in the crankcase sump to replace oil used in cylinder lubrication.

Box lubricators are driven either from the crankshaft, another moving part of the machine, or a separate electric motor. They contain a reservoir for oil and individual pumping elements. A camshaft operates the pumping elements. Two types of pumping elements are used in a box-type lubricator system: pumps with sight glasses and pumps with pressurized supplies.

Common Oil Supply
When process gas composition and cylinder operating conditions allow compressor frame lubricating oil to be used for cylinder and packing lubrication, a common-supply force-feed lube system is used (Figure 5.4).

Independent Force-Feed Lubricator Systems

The cylinders and packing require an oil supply at the lubricator pump inlet. Measures that may be necessary to make sure that the force-feed pump is filled with oil during the suction stroke include appropriate pipe and fitting size from the tank to the force-feed pump, heating the oil, and pressurizing the supply tank.

Independent Oil Supply
When process gas composition and cylinder operating conditions require an independent cylinder and packing oil supply, the resulting separate force-feed lube systems require an oil

supply. Lubricator oil is supplied under pressure from an elevated tank. To prevent the compressor frame oil from being contaminated by the force-feed oil, be sure that the lubricator box overflow does not drain into the crankcase. The lubricator box overflow tubing must be disconnected from the compressor frame and directed to an appropriate drain system (Figure 5.5).

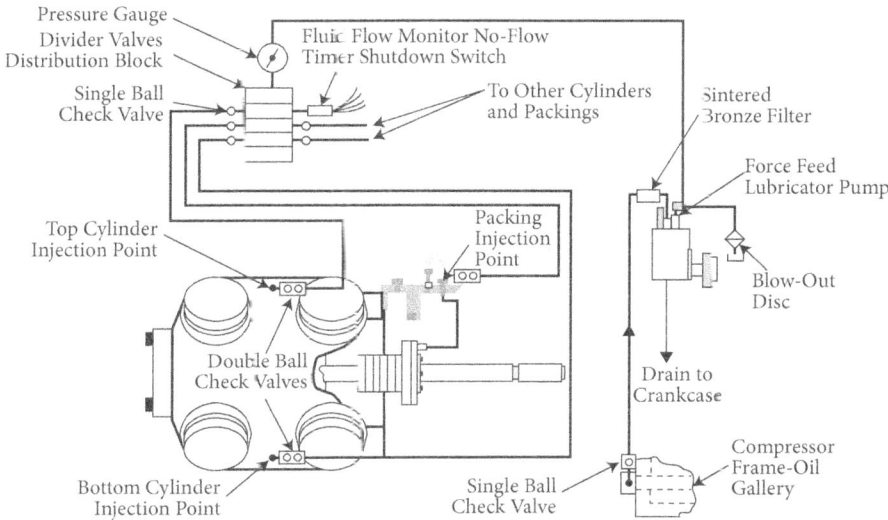

FIGURE 5.4   Common-supply force-feed lubrication system.

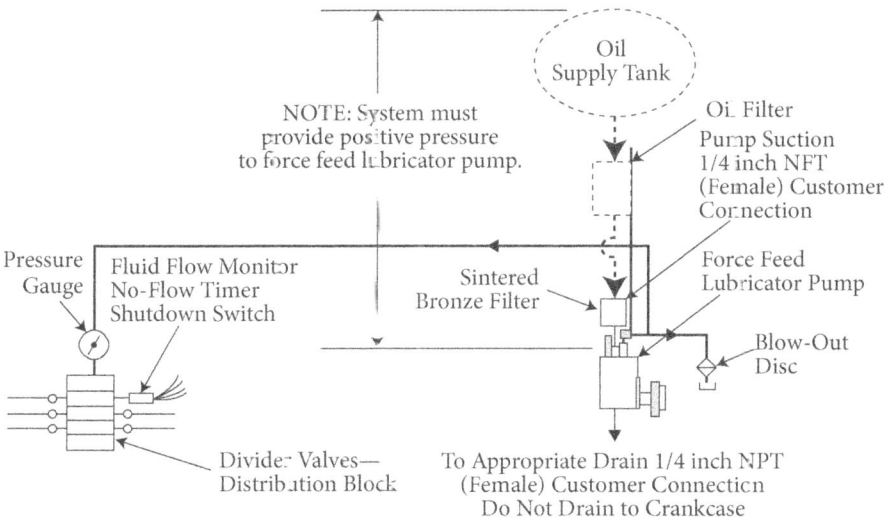

FIGURE 5.5   Independent-supply force-feed lubrication system.

Injection Oil Inlet Filter

An inline oil filter or fine screen is required between the oil supply and supply tank and the force-feed lubricator pumps. It is recommended to have filtration of 5 μm nominal. The compressor filtration system is adequate for systems that use frame lube oil for the force-feed cylinder and packing injection. For separate force-feed lube oil supplies, a filter must be installed by the packager.

Oil Dilution

Cylinder lubrication requirements will vary with the operating conditions and the composition of the gas to be compressed. Careful consideration must be given to proper cylinder lubrication selection. The degree of cylinder oil lubrication saturation by the process gas stream is influenced by the following factors:

1. **Process gas composition/specific gravity.** Usually the higher the specific gravity, the greater is the oil dilution.
2. **Discharge gas pressure.** The higher the pressure, the greater is the oil dilution.
3. **Discharge gas temperature.** The higher the cylinder discharge temperature, the less is the oil dilution.
4. **Lubricant selection.** Some types of oil are more prone to dilution than others.

Cylinder Oil Lubrication

Lubrication rates may vary for various gas compositions and various operating conditions. At higher pressures, lube oils with higher viscosities should be used. But when liquids are present in the gas, the most effective lubrication of cylinders and packing requires removal of the liquids before the gas enters the compressor. The use of higher-viscosity lubricants or specially compounded lubricants can compensate somewhat for the presence of liquids in the gas stream. If the recommended lubricants or flow rates do not appear to work adequately, flow rates and lubricant types may need to be changed.

Under/Overlubrication

Inadequate.   Underlubrication results in extremely rapid breakdown of pistons and packing ring materials. Black, gummy deposits that can be found in the distance piece packing case, cylinder, and valves are indicators of underlubrication.

Excessive.   Overlubrication can result in excessive oil carryover into the gas stream and increased quantities of deposits in the valves and gas passages. Valve parts breakage and packing failure may also be symptoms of overlubrication.

Inadequate Lubrication Symptoms

When observed symptoms indicate inadequate lubrication, first verify that the force-feed lubricator pumps are operating properly. Confirm that the distribution block cycle time matches the lube sheet or lubrication box information plate provided by original equipment manufacturer (OEM), and double-check that all tubing and fittings are tight and no leaks are present. Do not overlook the fittings inside the cylinder gas passages. Pressure test or replace divider valves to be sure that they are not bypassing.

Lubricator Cycle Time

The lubricant flow rates are generally so low that all the required flow to a lube point may be observed as a drip at a loosened supply fitting. The break-in and normal lube timing rates, which are stamped on the lubricator box information plate, are calculated according to the lube specifications, as given in the company's operation and maintenance manual, to match the gas and operating conditions. The lube sheets supplied by OEMs state gas conditions and list the base-rate multiplier at each lube point. If the compressor operating conditions change, the lubrication rates must be recalculated, and hardware changes may be necessary to the force-feed lubrication system.

Force-Feed Lubrication System Monitoring

Minimum Requirements

Many compressor vendors require the unit control panel to shut the unit down if the master lubes distribution block stops cycling while the compressor is running. The control panel is programmed to wait for a maximum time of three minutes after the distribution block stops cycling to issue the shutdown. This is a class B shutdown. A class B shutdown is defined as a shutdown that is not armed until a short time after the compressor is started, typically two minutes. This gives the divider blocks time to cycle the first time before the shutdown is activated.

Devices

Compressor vendors offer several devices that can be used to initiate the lube system shutdown. There are two basic types. Note that some devices have both functions built in.

Choosing a Lube System Monitor

When ordering a compressor, the packager must choose a lube system monitor. For further information about such devices, consult the vendor literature.

Break-in Rate

A force-feed lubrication pumps should be capable of delivering 150 percent minimum of the "normal" required lube rate for the break-in period (set as close as possible to twice the normal rate for 200 hours). Break-in rates should be maintained for 200 hours of operation of new equipment or when replacing packing or piston rings.

Factors Affecting Cylinder Lubrication

In compressor cylinders, operating temperature has an important role because of its effect on oil viscosity and oil oxidation, as well as on the formation of deposits. At higher temperatures, oil viscosity is reduced, and hence higher-viscosity oils are required to maintain adequate lubrication films.

Oil on discharge valves and valve chambers is heated when it comes in contact with hot metal surfaces and is continually swept by the heated gas after the compression cycle. This is a severe oxidizing condition, and the compressor oils oxidize to an extent that depends on the conditions to which they are exposed. The oxidation products formed are deposited mainly on the discharge valves, which are the hottest parts. The deposits on discharge valves may get struck in the valve seating area, leading to leakage of hot, high-pressure gas back into the cylinders. This high-temperature gas heats intake gas on the suction stroke, leading to high discharge temperatures.

The oil fed to compressor cylinders is subjected to oxidizing conditions as a result of high temperatures encountered when the oil leaves through the discharge valves, where temperatures are highest. Hence, to minimize the formation of deposits on the valves, oil fed to the cylinders should be kept to a minimum. It also helps to reduce excessive oil carryover to downstream equipment. The Compressed Air Institute suggests the use of 1 pint of oil for each 6 million square feet of area swept by the piston in air compressors. These are general recommended starting points and may need to be adjusted (up or down) based on gas and operating conditions in the compressor.

Moisture is an important factor because condensation occurs in the cylinders during idle periods when cylinders cool below the dew point of the air. The water thus formed tends to displace the oil films and come in contact with the metal surfaces, leading to rusting. The rust, when stuck between the surfaces of piston and cylinder, will result in excessive wear. Also, rust tends to promote oil oxidation and contribute to the formation of deposits. If this potential exists, the oils should be selected with good rust-inhibiting qualities and fortified with effective additives that will adhere to metal surfaces.

Rate of Oil Feed

The amount of oil fed to the compressor cylinders should be sufficient to provide lubrication and effectively seal the piston against leakage. Oil feeds above this amount are wasteful, cause oxidation, and tend to increase oil carryover to distribution lines.

Excessive Oil Feed.   All oil fed to the cylinders is subjected to oxidizing conditions. Under prolonged heating, even the best-quality compressor oils will oxidize to some extent. Therefore, feeding more oil than is actually needed results in increasing the amount of oxidation products formed. Because the highest temperatures are encountered on discharge valves and in discharge passages and most of the oil fed to the cylinders eventually leaves through the discharge valves, it is here that deposits tend to accumulate. To prevent or minimize trouble from deposits, an oil especially suitable for compressor service that permits using very low rates of oil feed should be used. Feed rates for compressor cylinder lubrication are typically shown in drops per minute.

Cylinder Oil Feed.   In the lubrication of double-acting compressor cylinders, one of the most important factors is the rate of oil feed. The likelihood of overlubrication is greater than that of supplying too little oil. Many problems associated with compressor operation can be overcome by preventing excessive lubrication. Proper control of the supply of oil to the cylinders is the most effective means of preventing the formation of objectionable deposits around valve ports, in ring grooves, and on cooler surfaces.

If the frame bearings and cylinder are using the same oil for lubrication, the oil is pumped from a reservoir where the oil is filtered and cooled. The oil is distributed to the bearings in the frame by an oil pump. The same oil is also used, by way of injectors, to lubricate and seal piston rings. In many cases, two different lubricants are used. The frame bearings do not require synthetics because of the moderate conditions. High-pressure, high-temperature conditions may require synthetics or compounded oils in the cylinder. Also, higher pressures require higher viscosities. A separate oil system is used to supply oil to the injectors for the cylinder. Too much oil in the cylinder can create problems, such as carbonizing the valves. It is better to underlubricate than to overlubricate. The injectors are set in drops per minute and have to be adjusted to the size of the drops to achieve quarts-per-day calculations. During commissioning and run-in of new compressors, this amount is usually doubled. Check with the OEM for its requirements.

In most conditions, frames are lubricated with R&O oil (usually an ISO VG of 150). In some cases, an OEM may recommend an ISO 100 oil. Cylinder lubrication is related to the type and pressure of the gas being compressed. Units that compress inert gases are the easiest to lubricate—with ISO 150 R&O oil under moderate pressures (<1,000 lb/in^2). As pressures increase to 5,000 lb/in^2, there's a corresponding increase in viscosity from ISO 150 to 680.

Contamination Concerns

Dirt and air are the two primary factors involved in most lube systems failures. The lubrication system is a hydraulic system, and like any other system of this type, dirt in the oil can cause serious damage to lubrication system components. Even if it does not cause immediate failure, it can greatly reduce the reliability of the lube system and compressor. Proper filtration and clean oil are necessities.

Purging Oil Lines and Components

Although not usually the cause of damage in a lubrication system, the presence of air in lube lines and components is often the cause of lubrication failure. Finding where the air lies is often difficult, and eliminating its source also can be tricky. Remember, although air cannot usually damage the components of a lube system, it certainly does not lubricate the compressor very well, so all air must be eliminated. Purging of air in lube systems after installation, maintenance, or testing is very important. The proper purge method is to begin at the source and move progressively downstream, verifying at the inlet and outlet of each component that all air and contaminants have been eliminated. When new tubing is run, care should be taken to avoid pumping dirt, particles of cut tubing, and other contaminants into the components. These can be very damaging. Pump a large quantity of oil through long tubing runs before connecting them. A large displacement portable hand pump works well for this purpose. The choice of a cylinder lubricant is affected by the properties of the gas to be handled. In this respect, gases usually fall into one of three classes: inert, hydrocarbon, and chemically active gases.

CONCLUSION

The lubrication system in reciprocating compressors must inject sufficient quantities of clean fluid to provide lubrication for the compressor's internal parts, such as pistons and cylinders, and to provide a positive seal between moving and stationary parts. Most compressors consume more oil than necessary, so operating costs could be significantly reduced by taking small steps. Cost reduction can be achieved with relatively simple steps to ascertain the correct type and quantity of oil being delivered to each lube point without changing the design or materials of the wearing components. Some proven methods can be implemented to ensure that the lube system is operating correctly to provide long-term, reliable compressor operation.

ICML Questions

1. In reciprocating compressors, normally there are independent systems for lubricating the compressor bearings, cylinder, packing, and crosshead. Which of the following is false?
   a. The frame oil system supplies oil to the bearings.
   b. The force-feed system supplies oil to the cylinder.
   c. The frame oil system supplies oil to the packing.
   d. The frame oil system supplies oil to crankshaft, connecting rods, and crosshead.

2. In reciprocating compressors, normally prelubrication is required because
   a. before startup, it takes time to lubricate components.
   b. passage of oil in the connecting rod drilled holes takes time.
   c. lube circulation maintains temperature.
   d. All of the above

3. In reciprocating compressors, operating temperature (normally 66°C) must be maintained
   a. to drive off water vapor.
   b. to maintain viscosity.
   c. for better lubrication.
   d. to make the lubricant thinner.

4. In reciprocating compressors, crankcase oil heaters are specified for outdoor compressors because
   a. they keep the oil at the required viscosity.
   b. they prevent condensation.
   c. they prevent corrosion.
   d. All of the above

5. In reciprocating compressors, force-feed lubrication system excessive oil supply to the cylinder and packing can cause
   a. fouling of valves.
   b. gumming of the packing.
   c. accumulation in the downstream piping system.
   d. All of the above

6. In reciprocating compressors, a sampling point should be installed
   a. between the oil pump and the filter.
   b. between filter and bearing.
   c. between the pump and the cooler.
   d. anywhere in the system.

7. Factor that affect crankcase lubrication include
   a. load.
   b. speed.
   c. temperature.
   d. All of the above

8. Oil oxidation is accompanied by
   a. a gradual increase in viscosity.
   b. no change in viscosity.
   c. a gradual decrease in viscosity.
   d. either an increase or a decrease in viscosity.

9. Oils in cylinder and packing lubrication are
   a. delivered in once-through systems.
   b. can be reused.
   c. go to the main oil tank from the cylinder and packing.
   d. go to force-feed system cylinders and packing.

10. To minimize the formation of deposits on valves, oil fed to cylinders
    a. should be kept to a minimum.
    b. should be kept to a maximum so that sticky material will be removed from the valve.
    c. has no impact on feed rate.
    d. depends on the downstream equipment.

11. During idle periods, sometimes cylinders cool below the dew point of the air. To avoid condensation in the cylinders,
    a. oils should be selected with good rust-inhibiting qualities.
    b. the cylinder should be heated from time to time.
    c. water in the cylinder jacket should be circulated all the time.
    d. All of the above

CHAPTER 6

Lubrication of Screw Compressors

BASIC OPERATING PRINCIPLES

A rotary-screw compressor is a positive-displacement machine that operates without the need for suction or discharge valves. Screw compressors are commonly used in a variety of process gas, process refrigeration, and natural gas applications, including individual wellhead boosters, low-pressure gathering systems, fuel gas compression, and vapor-recovery compression systems. They have been used on sweet and sour gas as well as acid gas applications with H_2S and/or CO_2 concentrations in excess of 80 percent. Screw compressors can be used on very light gases such as hydrogen and very heavy mole weight gases with specific gravities approaching 2. We have seen screw compressors become much more common for the unconventional gases because of their wide operating range and turndown capability, as well as the reduced maintenance costs. Figure 6.1 shows a typical cutaway sketch of a rotary-screw compressor.

Some of the major components include one set of male and female helically grooved rotors and a set of axial and radial bearings, all encased in a common housing. Related components in the compressor assembly include the intake air filter, an oil pump, and the inlet valve. The moving parts in a screw compressor are the male and female rotors. The male rotor is driven either directly by the motor or through step-up gears in the drive housing. The male rotor, which has four lobes, drives the female rotor, which has six lobes. The two rotors are helically grooved and mesh to compress inlet air in one stage of compression. The pulse-free air is delivered to the receiver or the inlet of a second compression stage (Figure 6.2).

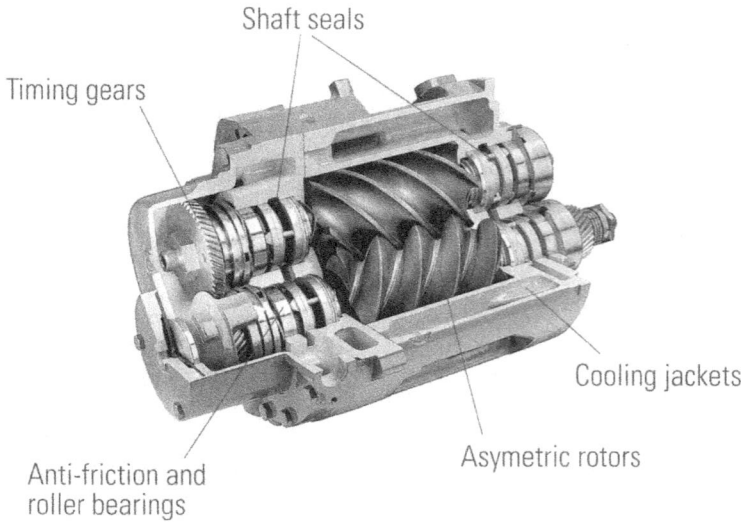

FIGURE 6.1   Typical cutaway sketch of a rotary-screw compressor.

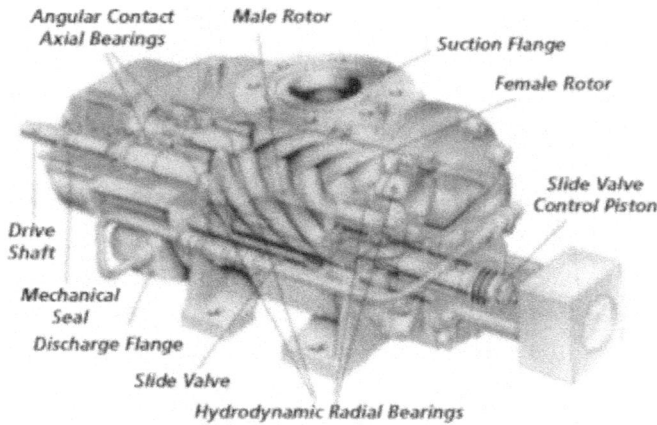

FIGURE 6.2   An oil-injected screw compressor.

Screw compressors used in the petroleum, petrochemical, refrigeration, and fuel gas industries tend to feature lubrication, shaft-sealing, and oil-control systems designed to the general requirements of American Petroleum Institute (API) Standard 614.

The first item to note is the difference between an oil-injected compressor's lubrication oil system and an oil-free compressor's lubrication oil system. The difference is that an oil-injected system features separation vessels that double as reservoirs (Figure 6.3), whereas an oil-free compressor has a more basic reservoir and is completely separate from the gas system.

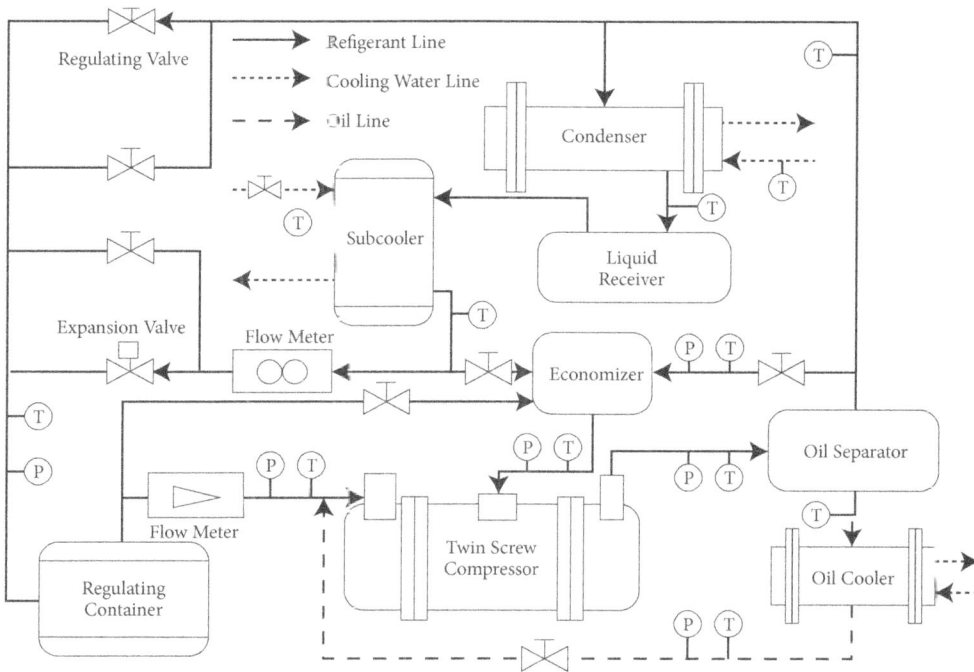

FIGURE 6.3 Piping and instrumentation diagram of a lubrication oil system for an oil-injected screw compressor.

Oil-flooded rotary-screw compressors require lube oil to provide sealing between the rotor lobes and the casing and the male and female lobes where the compression occurs. The oil is also required for lubrication of the bearings and shaft seals and to reduce the heat of compression in the machine. The lube oil system on a screw compressor is a closed-loop system. The oil is injected into the machine in several places. The main oil injection port feeds the rotors directly, with smaller lines feeding various points on the machine for seals and bearings. Once the oil is injected, passages within the machine will drain all the bearing and seal oil into the rotors, where it combines with the gas. The gas and oil mixture is then discharged out of the machine.

The gas becomes entrained within the oil, and the oil separator vessel is used to separate the two media. The level of the oil within the separator must be monitored because too high a level would increase the amount of oil carry-over into the process and too low a level could eventually cause a process trip.

This vessel can be either vertical or horizontal in design depending on equipment layouts and space availability. The vessel requires coalescing-type elements to remove as much oil as possible. Typical oil carry-over rates from the separator are in the 10-ppm range. The oil separator also acts as a reservoir for the lube oil system. The lube oil flows from the bottom of the separator through an oil cooler, where it is cooled from discharge temperature down to 140–160°F, through an oil filter and then back to the machine.

From the primary separator vessel, two oil pumps supply oil through separate pressure and temperature control systems before arriving at the oil filter. After the filter is a selection of valves that allow the oil to circulate freely or be injected directly into the compressor bearings and working chamber. If the compressor is running the oil return to the primary separator mixed with the process gas to be separated or if the compressor has stopped, any excess oil in the chamber can return via a drain valve.

The primary separator on oil-injected packages has two functions, one is to separate the gas and oil mixture that is discharged from the compressor and the other is to act as a reservoir for the oil pumps. The dimensions of the vessel, both diameter and height, are sized so that both functions can be carried out (Figures 6.4 and 6.5).

After the bulk of the oil has been knocked out of the gas using the diffuser, the process gas passes through a high-performance demister as the next part of the separation process. The demister is self-cleaning in operation, has a very small pressure drop (150 Pa), requires no maintenance, and can achieve separation levels down to about 100ppm by weight in isolation.

After the process gas exits the primary separator, some packages, depending on the application, have a secondary separator installed to further reduce the oil entrained in the gas to only a few parts per million. The oil collected by the secondary separator is returned to the lubrication oil system (Figure 6.6).

An oil reservoir that is the oil capacity of the primary separator will normally contain two minutes of retention capacity, where retention capacity is defined as the total volume below the minimum operating level. The minimum volume of oil in the vessel therefore must be two times the normal flow of oil to the compressor, where normal flow is defined as the total amount of fluid required by equipment components such as bearings, seals, couplings, and controls, excluding transient flow for controls or fluid bypassed directly back to the reservoir.

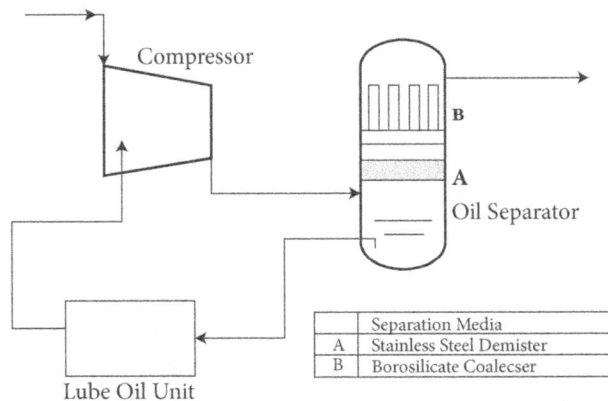

	Separation Media
A	Stainless Steel Demister
B	Borosilicate Coalecser

FIGURE 6.4   Oil separation arrangement: single-stage oil separator.

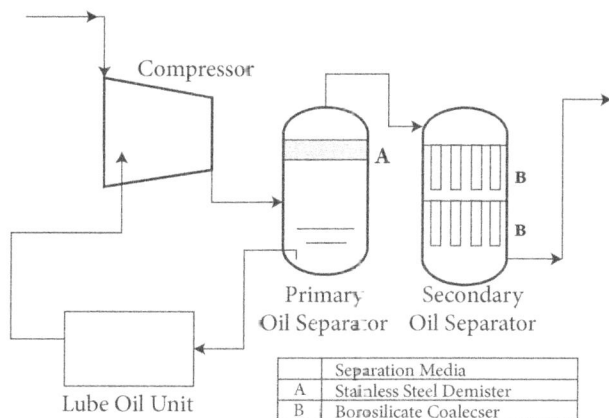

FIGURE 6.5   Oil separation arrangement: two-stage oil separator.

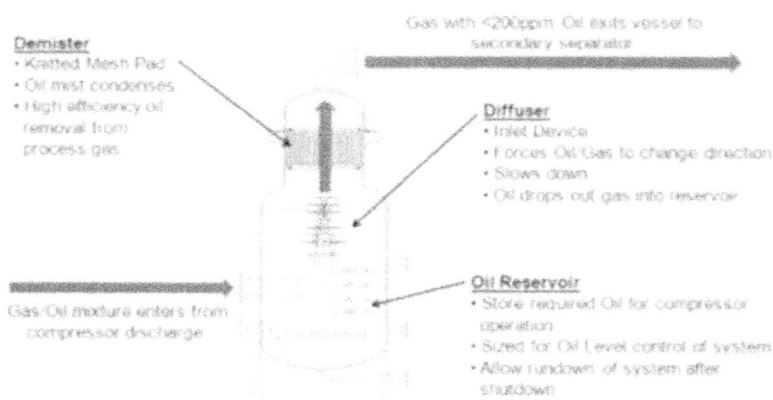

FIGURE 6.6   Sketch of a primary separator detailing the demister, diffuser, and reservoir.

The oil level within the separator at startup must be below the rotors so as to prevent the situation where the compressor is trying to startup and compress oil, resulting in a much increased starting torque. The compressor rotor shaft on the package will high enough to ensure that sufficient oil will drain from the compressor casing back to the separator. An oil level indicator is required on the separator. This could be a sight glass, level switch, or continuous-measurement transmitter. The range of the indicator should span from 50mm below the lowest operating level to 50mm above the rundown level, where the rundown level is defined as the highest level that the oil reservoir can reach when the entire system is shut down. When the system shuts down, the operating level while the compressor is running may have been at 60percent, but because all the oil in the system drains back to the reservoir after it has shut down, the rundown oil level may rise substantially.

It is important to understand that the oil level indicator, whether it is a gauge or a transmitter, is only used for alarm and indication: no tripping function is required to protect the machine. If the oil level falls such that the pump discharge pressure falls, the transmitter monitoring the oil-to-gas differential pressure would detect this and shut the package down.

Primary separators are fitted with oil heaters to prevent the compressor with oil so cold and viscous that it will not circulate properly. This is particularly important for packages in cold environments. The heater also prevents the buildup of liquid refrigerant (condensate) in the separator during shutdown. The heater should be fitted below the minimum oil level, have a safety high-temperature cutout, and have a maximum rating of $2W/cm^2$ to prevent burning the oil. The heater should be sized appropriately to heat the oil within the separator in 12 hours.

During normal operation, the oil level within the separator will gradually decrease as a result of the oil carryover. To prevent having to stop the system to top up the oil level, a complex piping arrangement featuring multiple nonreturn valves, top-up tank, and pump can be installed to facilitate replenishing the oil while the system is still running. The top-up pump discharge pressure must be higher than the compressor discharge pressure in order to force the oil into the separator. The starting and stopping of the top-up pump is a manual function that can be done via a local control station or the human-machine interface (HMI).

In order to begin the prestart sequence, certain conditions must first be met. The primary separator level must be above the low alarm limit, and the lubrication oil temperature within the separator must be above a certain predetermined value. When these parameters are satisfied, the lubrication oil system can be started to allow the system to get up to pressure and temperature prior to the compressor start. During the prestart sequence, the duty lubrication oil pump, pumps oil from the separator, through the oil-to-gas differential pressure control system, through the temperature control system, and through the filter before bypassing the compressor and returning to the separator via an oil recirculation valve and separate nozzle. Once the oil manifold temperature (different from the separator temperature) has maintained a steady 30–32°C, the lubrication oil system is ready for a compressor start. During the prestart sequence, the conditions within the primary separator are relatively static, with the oil still free of any process gas, still at a relatively low temperature, and entering the vessel from conventional nozzles.

During startup of the actual compressor, the drain valve is closed, the oil recirculation valve is closed, and the compressor manifold isolation valve is open, allowing oil to be sent to the compressor bearings and working chamber. After these valves have traveled and the oil-to-gas differential pressure control valve has reestablished a healthy pressure, the main drive motor of the compressor is started. During the startup sequence, the conditions within the primary separator are the same as the prestart conditions.

When the compressor is running, the conditions inside the primary separator can only be described as chaotic. The oil and gas mixture discharged from the compressor is blasted off the diffuser, sending it everywhere; the heat generated by the compressor sends the separator temperature soaring; and the demister pads are also dropping oil back into the reservoir before being sucked back out of the separator by the lubrication oil pump to go round again.

The stopping sequence, in terms of the lubrication oil system, is a reversal of the startup sequence: after the main driver has been stopped, the manifold isolation valve is closed, the recirculation valve is opened, and the drain valve is opened. This allows any oil within the chamber to drain back to the separator but also allows the oil to continue to circulate in the event that the system has to be restarted within a short time frame. During the prestart sequence, the conditions within the primary separator are relatively static with the oil still free of any process gas, still at a relatively low temperature, and entering the vessel from conventional nozzles. During the stopping sequence, the conditions within the primary separator become relatively static, similar to the prestart condition, with the oil still at a relatively low temperature and entering the vessel from conventional nozzles. The difference is that the oil is now entrained with process gas for a retention time, where the retention time is the time allowed for disengagement of entrained gas.

The density of the oil also changes within the separator between prestart and running. The large increase in temperature makes the oil expand, and because density is defined as the mass of fluid per unit volume, the mass stays the same, and the volume increases, thus reducing the density. In addition, as the process gas begins to be entrained in the oil, the density decreases further. The process gas composition also can be dynamic.

Depending on the composition of the gas stream in your system, it can dilute the compressor oil. The gas stream may consist of hydrocarbon gases such as propane, inert gases such as nitrogen, or aggressive gases such as methyl chloride, all in varying concentrations. Under pressure, many gases tend to be dissolved in certain lubricating oils and may reduce their viscosity significantly.

Lubricant additives or base oils may react with components of the gas, leading to the formation of sludge and deposits. Premature wear, extra downtime for maintenance, high oil consumption, and shorter oil change intervals are common consequences. With the right lubricating oil, you will not only reduce wear and increase the life of individual components, as well as that of your machines as a whole, but you will also boost your operational reliability and hence your profits. But which lubricating oil will release this potential?

SELECTION OF LUBRICATING OIL

Viscosity and Gas Solubility

An important factor especially for the lubrication of bearings and rotors is the right viscosity of the lubricating oil, which under all operating conditions should comply with the instructions provided by the compressor manufacturer. If the gas stream did not affect the lubricating oil, the choice of viscosity would be fairly simple. But certain gases tend to be dissolved in certain lubricants. This effect may considerably reduce the oil's viscosity and prevent the formation of an effective lubricant film. This, in turn, could lead to reduced service life or failure of the "air end" and hence machine downtime.

The composition of the gas streams to be compressed is as varied as the processes for which the compressors are used. The lubricating oil for the compressor should be chosen just as individually. Every gas stream is unique. One can check whether the individual gases have a tendency to dissolve in the lubricating oil.

Important factors for selection of lubricating oil include:

- **Pressure.** The higher the compression, the higher is the tendency of the gas to be dissolved in the oil.
- **Temperature.** The higher the compression temperature, the lower is the tendency of the gas to be dissolved in the oil.
- **Molecular weight.** The higher the molecular weight of the gas, the higher is its tendency to be dissolved in the oil. Heavy hydrocarbons such as toluene are more easily dissolved than light ones such as methane or propane.
- **Polarity.** When it comes to polarity, the principle of likeness applies. Polar gases such as ammonia are easily dissolved in polar oils such as polyglycol, whereas homopolar gases such as hydrocarbons are less easily dissolved in polar oils.

Taking into account these criteria, one can calculate the tendency of a gas to be dissolved in the lubricating oil already before the rotary-screw compressor is filled with lubricant. Paying attention to these criteria provides valuable information about the oil's viscosity under actual operating conditions and can suggest a lubricant that will meet the manufacturer's viscosity requirements in all real-life situations.

Selection of Oil

For dry-screw compressors, lubrication is required for the gears and bearings only. Premium rust- and oxidation-inhibited oils of a viscosity suitable for the gears are usually used. In some cases, oils with enhanced antiwear characteristics may be desirable for added protection of the gears.

For flooded-screw compressors, lubricating oils of ISO VG 32 (SAE 10) to ISO VG 68 (SAE 20) are used for plant air applications. Sometimes oils with a high level of detergents and corrosion inhibitor are used to pick up moisture. In low ambient temperatures, cyclic operation, or very humid conditions, rust- and oxidation-inhibiting oils with good demulsibility characteristics should be used.

When problems with varnish and sludge can be encountered in severe operating conditions, synthetic oils are replacing mineral oil. PAO-based synthetic lubricants are now the most commonly used oils in these applications. Other types of synthetic lubricants, such as diester, polyglycol, and synthetic blends also can be used. These materials also allow the extension of change intervals.

In refinery applications, mixing of the oil with the hydrocarbon gas occurs, for which oil selection is more critical than in air applications. In handling hydrocarbon gases, polyglycol-based synthetic lubricants are used. The polyglycols have low solubility for hydrocarbons, and thus dilution of the lubricant is minimized and good separation of the oil and gas can be obtained.

Problems with Residues, Deposits, and Sludge in the Oil

When mixing under high pressure and temperature, gases not only may be dissolved in the oil, but they also may react with base oils or additives. Conventional mineral oils, in particular, contain unsaturated hydrocarbons and sulfur compounds that tend to react with certain gas components. Furthermore, base oils and additives may be affected by acids or water vapor contained in the gas stream. The complex chemical system present in a compressor hosts a vast potential for chemical reaction and polymerization. A typical consequence of this type of reaction is the formation of sludge or solid residues. Both may severely affect the lubricity of the oil and lead to considerable wear on rotors and bearings.

Another possible effect is clogging of oil filters and separators, which, in turn, leads to higher oil consumption. In such a situation, high consumption of oil filters and separators, short oil change intervals, frequent machine stoppage for maintenance, and high oil consumption—all involving substantial costs—are inevitable. Chemical compatibility may be the cause. To avoid the formation of reaction by-products and prevent avoidable costs for maintenance and spare parts, it is necessary to check the compatibility of the gas stream and the lubricating oil before that oil is used in a particular compressor.

An accurate analysis of the gas stream being compressed enables the lube oil manufacturer to find a suitable lubricant for rotary-screw compressors that will result in fewer reaction by-products, deposits, and sludge; lower filter consumption; lower costs owing to longer oil separator life; lower oil consumption from longer oil change intervals and high separator performance; less wear on rotors and bearings (and consequently longer rotor and bearing life);

and less downtime for normal and extraordinary maintenance stops to replace oil filters, oil separators, bearings, and rotors.

High Oil Consumption and High Oil Carryover to the Gas Stream

Lubricants for the compression of gases such as pure hydrogen or nitrogen must keep the gas stream pure. At high compression temperatures, the oil injected into the compressor tends to evaporate—in particular, standard-quality mineral oils—and is carried along with the gas stream. There it is present in two forms: as oil vapor and as oil aerosol (small droplets). The aerosol is mostly filtered out mechanically by the oil separator; the oil vapor, however, can cause problems because it remains in the gas stream and is one of the causes of continuous oil consumption. Consequently, the more oil evaporates, the more is consumed. In chemical processes where the purity of the gas stream is essential, the presence of oil in the gas is highly undesirable.

Moreover, a high oil content also can cause problems in applications in the oil and gas industry. Downstream catalysts may suffer damage because of the effect of certain oils and additives. Components such as heat exchangers and pipes may be clogged by condensing oil vapor. In addition, high oil content in the gas stream leads to high oil consumption.

CONCLUSION

In screw compressors, a mixture of process gas and lubricant comes in contact with compressor bearings. The buildup of the lubricant film is governed by the amount and state of the mixture local to the bearings, as well as by the pressure, temperature, and flow transients in and around the bearings. This means that not only the local pressure and temperature are important to lubrication but also what happens to the process gas-lubricant mixture upstream of the bearing position and how fast it moves through the tubing compared with the time needed for the process gas to boil off from the lubricant. Viewing the process gases themselves as contaminants dissolved in the lubricant, the only detrimental effect on lubrication of most of them is to lower the viscosity and viscosity-pressure coefficient, leading to oil films that are too thin. Hence proper lubricating oil selection for the application is important to enhance the reliability of the system.

ICML Questions

1. Screw compressors
   a. are commonly used with a variety of process gases.
   b. have high turndown capability.
   c. have very high-flow applications.
   d. a and b only

2. Which is true for an oil-injected compressor's lubrication oil system and an oil-free compressor's lubrication oil system?
   a. An oil-injected system features separation vessels that double as reservoirs.
   b. An oil-free compressor has a more basic reservoir and is completely separate from the gas system.
   c. Oil-flooded rotary-screw compressors require lube oil to provide sealing between the rotor lobes and the casing and the male and female lobes where the compression occurs.
   d. All of the above.

3. Important factors for selection of lubricating oil in a screw compressor is (are)
   a. Pressure and temperature.
   b. Molecular weight.
   c. Polarity.
   d. All of the above

4. When problems with varnish and sludge are encountered in severe operating conditions in a screw compressor, which oil is best suitable?
   a. PAO-based synthetic lubricant.
   b. Mineral oil.
   c. Gaseous lubricant.
   d. Grease.

CHAPTER 7

Lubrication Systems for Turbo Machinery

Turbo compressors require lubrication to either cool, seal, or lubricate internal components. A properly designed, installed, operated, and maintained lubrication system is necessary for reliable turbo machinery performance. The most common causes for reduced reliability of these machines is the lubrication and seal oil system.

The lube oil system for a turbo machine must provide the rotating equipment with the necessary amount of:

- The correct fluid
- At the correct pressure
- At the correct time

The proper amount of fluid is needed to remove bearing and seal losses and control the temperature of these components to reduce wear and maintain proper clearances. The correct fluid is that for which the rotating equipment is designed to run and that which the oil console is designed to deliver at the required flow, pressure, temperature, and cleanliness. The use of other than the design lubricant may cause or contribute to problems such as premature wear, rotor instability or vibration, and poor system response. The correct pressure ensures adequate flow through bearing orifices for cooling and seals for cooling and sealing. The correct flow and pressure must be available when required to avoid process trips, nuisance alarms, bearing and seal failures, and safe operation.

These factors become more significant for larger and more complex systems.

However, these systems receive less attention after installation and become the focus of plant managers only when a spurious trip occurs. In this chapter, the principles involved in the design, operation, and maintenance of a turbo machinery lubrication system are emphasized.

The primary components of a system are described, including the design aspects. The chapter also reviews a number of problems arising in the lubrication systems of turbo machinery and provides examples from practical experience.

CIRCULATING LUBE OIL SYSTEM COMPONENTS

The lube oil starts off in the system reservoir. From there, it is drawn by pumps and fed under pressure through coolers and filters to the bearings. On leaving the bearings, the oil drains back to the reservoir. A circulating lube oil system is shown in Figure 7.1.

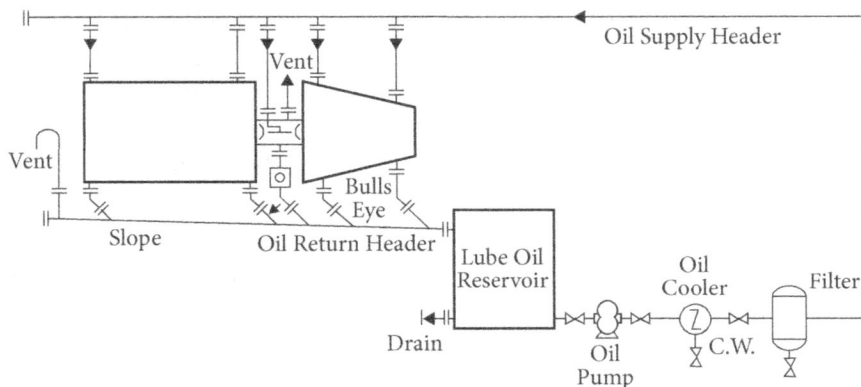

FIGURE 7.1   Circulating lube oil system schematic.

When in operation, the compressor lubricating oil is normally circulated by the main oil pump. An auxiliary pump serves as a standby. These two pumps generally have different types of drives or power sources. When both are driven electrically, they are connected to separate supply feeders. Heat generated by friction in the bearings is transferred to the cooling medium in the oil coolers. A pressure-regulating valve is controlled by the pressure downstream of the filters and maintains constant oil pressure by regulating the quantity of bypassed oil. A pressure switch activates the auxiliary oil pump. If the oil pressure falls below a preset limit, a second pressure switch shuts down the compressor train.

The flow of oil to each bearing is regulated individually by orifices, which is particularly important for lubrication points requiring different pressures. Temperatures and pressures are measured at all important locations in the system, including temperatures from oil sumps and return lines from bearings, gears, and other mechanical components (Figure 7.2).

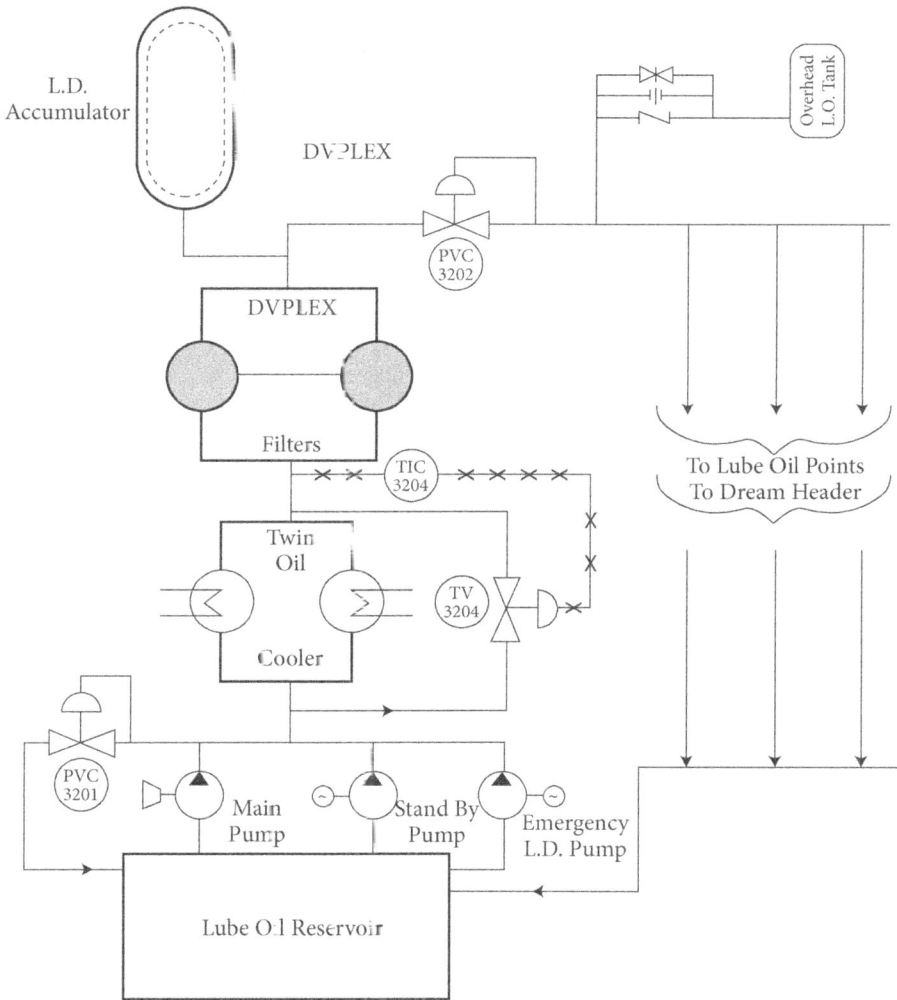

FIGURE 7.2   Simplified sketch of a circulating lube oil system.

The major components of force-feed lubrication systems in turbo machinery include the lubricant/oil, filter subsystem, cooling subsystem, oil pumps, vessel and piping, console/skid, and instrumentation/alarm.

Console/Skid

Most of the lubrication system components (pumps, vessel, etc.) are conveniently installed in a packaged unit called *lube oil console*, supplied complete and ready for installation. Oil pumps, coolers, and filters are grouped around the oil reservoir on a common base plate.

Design and construction of the lube oil system takes into account relevant regulations and any special requirements. Because auxiliary equipment must be maintained and calibrated during operation, it is important for the console/skid to be sized with enough space for maintenance personnel.

The reservoir is provided with the necessary connections for filling, draining, venting, and inspecting and oil purification. A portable oil conditioner is provided for purification of the oil (Figure 7.3).

FIGURE 7.3   Typical modular oil console/skid arrangement.

Vessel and Piping

The vessel functions as the oil reservoir for the system. The correct sizing is critical. Size will be a function of system flow and subsequently the amount of flow the hydro equipment (main guide bearings, thrust bearings) will actually pass. The function of the piping is to connect the console/skid auxiliary equipment (pumps, vessel, etc.) to the hydro units it services. The typical oil velocities are on the order of 4 to 6 ft/s.

Oil Pumps

The function of the oil pumps is to continuously supply the system with lubricant at the required flow rate. This means that it must be capable of uninterrupted operation for the same period as the turbo machinery it is servicing. Oil is normally circulated by the main oil pump. An auxiliary pump serves as standby.

Each pump is capable of supplying the capacity required by the whole stream. Therefore, each pump must be able to operate continuously. The standby pump starts automatically by means of a pressure switch in the common discharge of both pumps.

The suction line of each pump is provided with a strainer and isolation valve and pressure gauge. Pump running indication is provided via a pressure switch in the pump discharge. The two discharge lines are joined into a single connection line leading to the coolers.

These two pumps generally have different types of drives or power supplies. When both are driven electrically, they are connected to separate supply networks. On compressors with step-up gearboxes, the main oil pump may be driven mechanically from the gearbox, and the auxiliary pump then operates during the startup and rundown phases of the compressor train. Relief valves protect both pumps from the effects of excessively high pressures. Check valves prevent reverse flow of oil through the stationary pump.

Filter Subsystem

The function of the filter subsystem is to continuously provide clean auxiliary fluid (oil) to the critical equipment. A typical filtration specification for an auxiliary system is 10 absolute particle size, meaning the greatest size of any solid particle in the oil film should be 10 μm. There are two types of filtration systems: inline and offline filtration. The *inline* filter subsystem consists of a transfer valve, which allows transfer from one bank of components to the standby bank of components without significant pressure pulsations being introduced into the system, filters, differential pressure indicator, and alarm. *Offline* filtration, often called *kidney loop filtration*, functions independently of the designed lubrication system of the unit.

The filters clean the lube oil before it reaches the lubrication points. A differential pressure gauge monitors the degree of fouling of the filters.

Cooling Subsystem

The function of this subsystem is to continuously provide cool auxiliary fluid (oil) at the required temperature to the turbo machinery. Heat generated by friction in the bearings is transferred to the cooling medium in the oil coolers. The return temperature is monitored by a temperature switch. The oil is cooled by two water coolers, one as standby to the other one. Each cooling element is designed to cool the whole oil flow necessary for the system.

The oil flow is directed to one of the coolers, leaving the other one free for cleaning. This is realized by means of two three-way transfer valves that are simultaneously controlled by a hand lever. To equalize the oil pressure and facilitate the changeover operation, the two coolers are connected by a bypass pipe incorporating a needle valve. Air-cooled oil coolers may be employed as an alternative to water as a coolant in regions where water is in short supply.

Overhead Oil Tank

An overhead oil tank can be provided to ensure a supply of lubricant to the bearings in the event of faults while the compressor is being run down. A continuous flow of oil through an orifice maintains the header oil constantly at operating temperature. If the pressure in the lube oil system falls, the nonreturn valve beneath the tank opens to provide a flow of oil.

Instrumentation/Alarms

The function of the instrumentation is to measure and regulate the process variables of the auxiliary fluid (oil) such as flow, temperature, level, and pressure. Pressure indicators, temperature indicators, and differential pressure transmitters are examples of key instrumentation.

Temperatures and pressures are measured at all important locations in the system; the readings can be taken locally or transmitted to a monitoring station. The flow of oil to each bearing is regulated individually by orifices, particularly important for lubrication points requiring different pressures. Lube oil for the driver and other users is taken from branch lines. A pressure-regulating valve is controlled via the pressure downstream of the filters and maintains constant oil pressure by regulating the quantity of bypassed oil. The auxiliary oil pump is switched on by a pressure switch if the oil pressure falls. A second pressure switch shuts down the compressor train if the pressure still continues to fall.

Lube Oil Pressure Control Valves

The lube oil pump discharge pressure control valve (PCV) maintains a constant pressure [~8.5 kg/cm (gauge) normally in medium-sized machines] by bypassing the excess oil being pumped back to the reservoir. The valve settings are accomplished with the oil pump in operation and with stop valves and bypass valve operating. The pressure gauge in the line in which valves maintain the required pressure slowly closes the bypass valve and at the same time sets the control valve as it reaches the required pressure in the header when the bypass valve is fully closed.

There is another bearing lube oil PCV. This lubricating oil PCV maintains a constant pressure of 2.5 kg/cm (gauge) in the lubricating oil header. The lube oil pressure to bearings is adjusted by means of the flow regulator provided in each oil inlet. We provide a pressure of approximately 0.9–1.3 bar (gauge) [12–18 lb/ft^2 (gauge)] to journal bearings and approximately 0.3–1.3 bar (gauge) [5–18 lb/ft^2 (gauge)] to thrust bearing. The regular lube oil flow through the bearings can be checked by means of flow glasses.

SEAL OIL SYSTEM

The seal oil system in turbo compressors supplies mechanical contact and floating-ring seals with an adequate flow of seal liquid at all times, thus ensuring that they function correctly. An effective seal is provided at the settle-out pressure when the compressor is not running. The seal oil system may be combined with the lube oil system if the gas does not adversely affect the lubricating qualities of the oil or provided that the oil made unserviceable by the gas does not return to the oil system.

There are two methods of combining lube oil and seal oil systems: booster or combined systems. In the *booster* system, the oil pressure is raised to the pressure required for lubrication purposes, and then part of it is raised further to the pressure needed for sealing. Alternatively, in the *combined* system, all the oil is initially raised to the required pressure and flow and then reduced to the requirements of system component requirements.

Mechanical face seals and floating-ring seals are supplied with seal oil at a defined differential pressure above the reference gas pressure (pressure within the inner seal drain). The flow of seal oil is regulated by a differential pressure regulating valve that regulates the pressure of the seal oil if there are changes in the reference gas pressure or by a level-control valve that maintains a constant level in the overhead tank. The oil in the overhead tank is in contact with the reference gas via a separate line. The static head provides the required pressure differential. In addition, the oil in the overhead tank compensates for pressure fluctuations and serves as a rundown supply if pressure is lost. If the level in the tank falls excessively, a level switch shuts the compressor plant down. There is a constant flow of oil through the overhead tank, and this heats the oil at all times.

To prevent the oil from gaining ingress to the compressor, the space between the oil drain and compression space is sealed by a flow of gas. The pressure of this sealing gas is above the pressure of the reference gas. A differential pressure indicator monitors the pressure differential.

Seal oil is controlled at some specified differential pressure above the reference pressure of the medium being contained (usually gaseous). This differential pressure varies depending on specific seal designs and manufacturers. The soundness of the oil seals and seal oil delivery system for safe, reliable, and positive shaft sealing has resulted in extending the seal application to other machinery.

A schematic representation of lube oil and seal oil system is shown in Figure 7.4. There is a wide variety of oil systems in service. However, the basic system just described facilitates the discussion that follows.

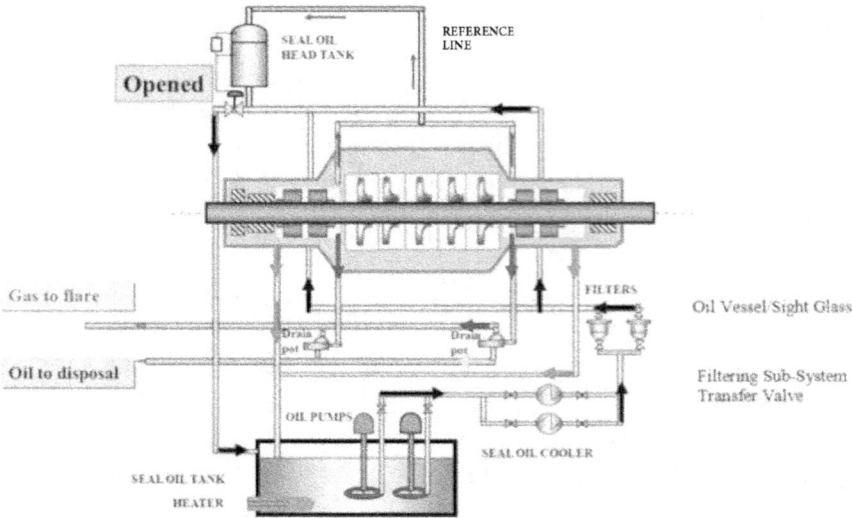

FIGURE 7.4   Basic seal oil system.

The different components of turbomachinery lube oil and seal oil system, their design considerations, and problem areas during operation are described next.

Reservoir

Purpose

The reservoir provides a quiet place in the circulating system where contaminants, water, and vapors that have entered into the oil in the bearing housings and other contact areas can separate out. The oil reservoir helps to dissipate or settle contaminants, air is dissipated via proper baffling and adequate residence time, and particulate matter is allowed to settle in the low end of the reservoir. Residence time and flow rates in the reservoir determine particulate deposition in the reservoir. Water is heavier than oil. Hence the low end of the reservoir must be designed for water drainage. The reservoir also helps in temperature fluctuations and expansion volumes and is the location for heating and oil-purifier connections. It stores a prescribed amount of oil and provides for rundown capacities.

Capacity

Factors involved in reservoir size and selection include:

- System flow rate
- Retention time

- Working capacity
- Height of return line from process equipment
- Rundown capacity
- Location of auxiliary pump

Reservoir capacities are in accordance with API 614 and are based on normal oil system flows. Figure 7.5 shows a schematic of an API 614 oil reservoir listing oil reservoir capacities.

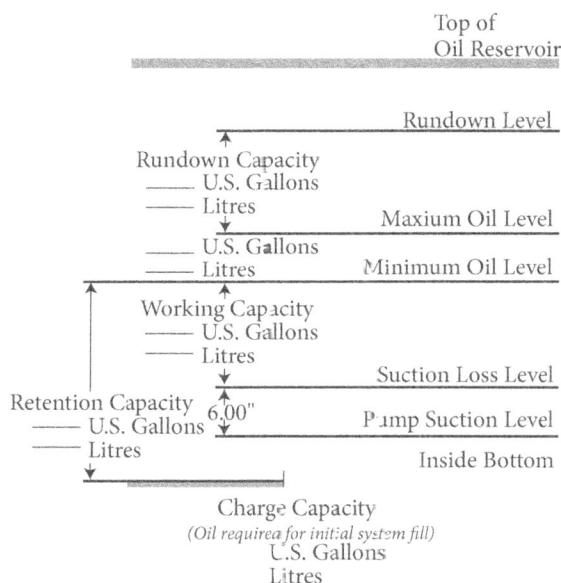

FIGURE 7.5 Oil reservoir capacities (schematic).

Normal Flow

Normal flow is defined as the total amount of oil flow that is required at the bearings, seals, couplings, and steady-state controls. Normal oil flow does not include oil that is bypassed directly back to the reservoir or transients. Based on this definition, oil reservoirs are designed in accordance with the following:

Working capacity. This is the volume of oil between the minimum operating level and pump suction loss level (level at which the oil pump loses prime—typically 2–6 inches above the pump suction level). This capacity is sufficient for five minutes of normal oil flow.

Retention capacity. This is the total volume of oil below the minimum operating level. It is sufficient for eight minutes of normal oil flow.

Rundown capacity. This includes oil contained in all components, bearing and seal housings, control elements, and vendor-furnished piping that drains back to the reservoir. Also included is a 10 percent minimum allowance for the user's interconnecting piping. An additional amount must be included as an allowance for major interconnecting piping runs in addition to the 10 percent minimum because users usually fail to identify the need quantitatively. It is the highest level that the oil may reach in the reservoir. This happens typically after the unit has been shutdown, and oil from drain lines and overhead oil-containing vessels has emptied back to the reservoir. A minimum of 1 inch of freeboard (airspace) should be provided.

Free surface. API 614 requires a minimum free-surface area of 0.25 ft²/gal per minute of normal oil flow. This requirement may determine the configuration and proportions of the reservoir. For example, a 10- × 6-ft oil tank has a free surface of 60 ft², or the equivalent flow of 240 gal/min. As oil flows increase, and thus tank size, a rectangular tank becomes unattractive because of bottom and side stiffening requirements. Attention is then given to either horizontal or vertical cylindrical reservoirs. Based on the free-surface requirements just stated and plant layout or space limitations, the optimal configuration can be chosen.

Design Considerations

API 614 specifies the mechanical construction of the oil reservoir, its basic features, and standard and optional ancillary devices. The most relevant considerations in reservoir design include:

- Appropriate materials for the jobsite conditions
- Adequate rundown capacity
- Sources of overpressure, venting, and purging
- Accurate specification of ambient conditions and utility availability for heaters
- Adequate pump suction conditions

Material.   The reservoir material should be either carbon steel, stainless steel, or a combination of the two depending on customer requirements. If carbon steel is specified, a protective internal coating should be applied to prevent corrosion as a result of moisture from atmospheric condensation during shipment and operation.

Capacity and Residence Time

1. The normal operating volume of the reservoir should provide a total residence time of not less than eight minutes.
2. The required residence time in industrial applications may vary from as low as three minutes for low-viscosity oils to as high as 40–60 minutes for high-viscosity oils.

3. Higher-viscosity oils have higher specific gravities. Water separation becomes more difficult as the specific gravity of the oil approaches that of water. Increasing the temperature in the reservoir will increase the specific gravity differential between the oil and water, making separation more effective.

Size and the details of construction vary markedly from user to user for API 614 systems.

A typical rectangular tank schematic is shown Figure 7.6. Various oil levels are shown in Figure 7.5.

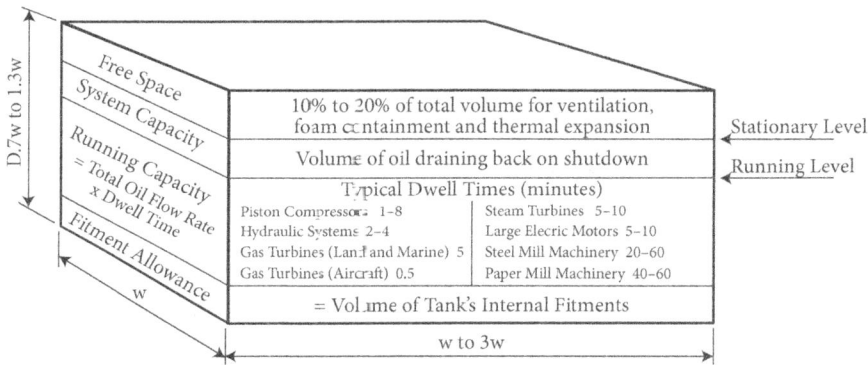

10% to 20% of total volume for ventilation, foam containment and thermal expansion	Stationary Level
Volume of oil draining back on shutdown	Running Level
Typical Dwell Times (minutes)	
Piston Compressors 1–8 Steam Turbines 5–10	
Hydraulic Systems 2–4 Large Elecric Motors 5–10	
Gas Turbines (Land and Marine) 5 Steel Mill Machinery 20–60	
Gas Turbines (Aircraft) 0.5 Paper Mill Machinery 40–60	
= Volume of Tank's Internal Fitments	

D.7w to 1.3w

*Free Space
System Capacity
Running Capacity
= Total Oil Flow Rate
x Dwell Time
Fitment Allowance*

w

w to 3w

FIGURE 7.6   Typical rectangular tank schematic.

Layout.   The layout of the reservoir is as follows. The reservoir should be located off the ground to permit drainage of water. The bottom of the tank should be sloped ¼ in/ft. Separation can be effective only if the reservoir is designed and sized properly. Oil should enter near the top and exit near the bottom of opposite ends of the reservoir.

Manway.   A gasket manway and riser should be supplied so that access is available to all compartments of the reservoir for inspection and cleaning. Reservoirs should be provided with a vent and an appropriate gauging device.

Pump Suction Lines.   Pump suctions are located at the high side of the slope in the tank. Pump suction lines should be straight pipe with a minimum number of elbows to avoid accumulation of air so as to result in smooth pump transfers. Avoid using elbows in the suction lines to prevent air pockets and minimize the potential for cavitation.

Return Lines.   All return lines and reservoir drains are located on the low side of the slope, to allow for the settling of dirt and water. Return lines should terminate below the oil level to minimize splashing and prevent foaming and electrostatic charge buildup. Return lines should be equipped with end baffles and diffusers, or they should be angle cut at 45 degrees. Return lines should discharge away from the pump suction and the reservoir bottom. A vent is required at the top of each return line.

Vent.   For lube oil and seal oil systems, tank overpressurization protection must be provided. For many situations, an oversized reservoir vent will be used. Depending on the size of the system, interconnecting piping layout, and pressure, these vents vary from 3 to 10 inches in size for rectangular API tanks.

Baffling arrangements and dimensions of reservoirs have varied according to individual requirements and space limitations. An ideal design is one in which the linear velocity of the oil is 1 ft/min maximum. Excellent separation of water can be accomplished at this velocity.

Baffles.   API 614 oil reservoirs are manufactured with a baffle plate running practically the total vertical distance of the reservoir, located at approximately the horizontal midpoint. The function of the baffle is to separate pump suctions from oil returns. All oil return lines enter the reservoir on the opposite side of the baffle from the pump suction lines to avoid disturbance or turbulence at the pump suction. Oil returns include returns from process pumps and auxiliaries, lube oil pump relief valves, PCVs, filter vents, control valve head, and instrument vents. Baffles should have an air passage at the top of the tank equal to three times the area of the auxiliary and main pump suction lines combined. A baffled passage is strategically located 2 inches above the tank bottom; this baffle should have the greatest possible distance between the oil pump inlet and return line to allow oil to pass from the return side to the suction side of the reservoir (Figure 7.7). Purge connections will be located on both sides of the internal baffle.

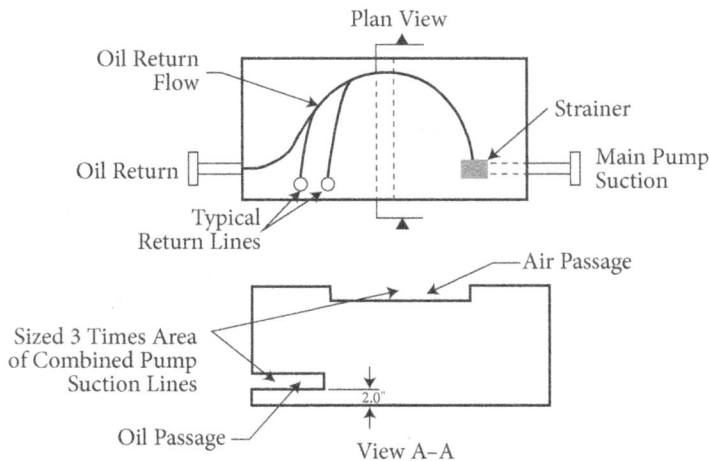

FIGURE 7.7   Reservoir baffle design and orientation.

Gravity drain return lines must extend 2 inches below the suction loss level. Drains with flows that exceed velocities of 3 ft/s (typically backpressure regulator and relief-valve returns) must extend 4 inches below the minimum operating level. In both cases, the oil is admitted into the reservoir through vented stilling tubes terminating into diffusion plates.

Pump suction lines are located on the opposite side of the reservoir baffle. There will be no additional drains on this side. Pump suction piping must be sized to maintain proper pump suction conditions, particularly when suction strainers are used. Low velocities, adequate tank head, and antiswirl plates virtually eliminate vortexing.

Reservoir Fill Cap. The filler breathe cap should be located on a riser to prevent water from running into the reservoir. The fill cap should have a 40-μm breathing element and a 60 mesh strainer to prevent foreign airborne particles and objects found in new drum oil from being introduced into the system.

Additional Items Required for a Reliable Design

Armored Sight Glass. A weld pad sight glass should be added to the reservoir. An armored reflex glass is required. The sight glass should span an area from 1 inch above the rundown level to 1 inch below the minimum operating level of the of reservoir (Figure 7.8).

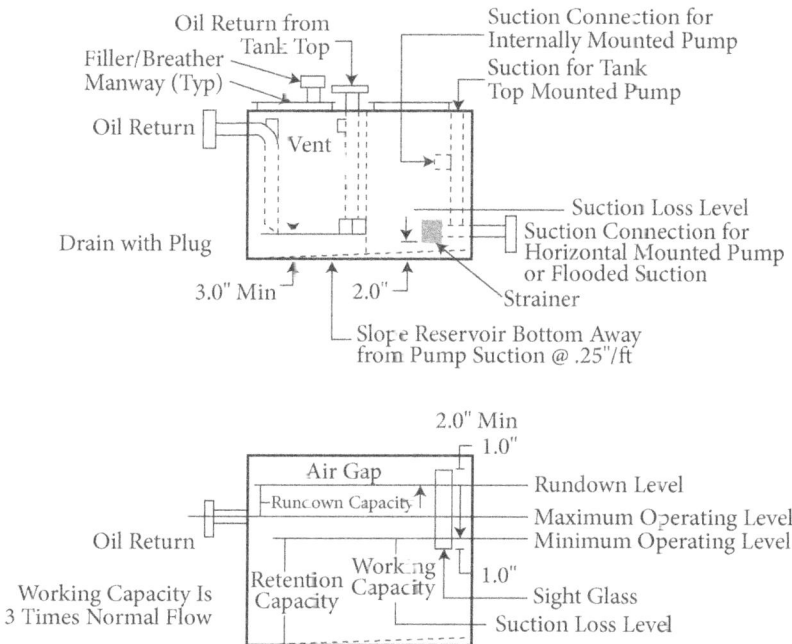

FIGURE 7.8   Reservoir suction and return line orientation/reservoir capacities/sight glass installation.

Oil Purification Connections.   To facilitate oil cleanup, it is recommended that oil purification connections be provided with isolation valves to the reservoir. The suction connection has a gooseneck with a 0.25-inch siphon breaker hole at the minimum rundown level. The oil return terminates below the oil level and includes a stilling tube. This option allows the user to connect a remote oil purifications system.

Potential Reservoir Problem Areas

Improper Reservoir Venting.   To protect the reservoir from overpressurization in the event of a seal failure, the reservoir vent area should be sized to handle the predicted pressurization rate. Blow-off covers, ruptured disks, and combination relief valves and flame arrestors are used, but typically, oversized atmospheric vents are used. Units with nitrogen purging are typical offenders of this requirement.

Electric Heater Short Circuits.   Certain heater manufacturers use a ceramic insulator in their heaters. The heater element is installed in a heater sheath supplied in the reservoir. During shipment, the ceramic is susceptible to moisture and therefore can later short out. The insulator should be baked in an oven prior to installation. If shorting out occurs during normal operation, the wiring conduit should be checked to ascertain that it is properly sealed and watertight.

Carbonizing Electric Heaters.   Care should be exercised to ensure that the heater is totally submerged prior to energizing. On oil reservoirs, the heater is installed a minimum of 1 inch below the pump section.

Most of these heater problems have been experienced in degassing tank applications, where low seal leakage rates have resulted in an extended time required to fill the tank to the operating level. Prefilling the degassing tank has eliminated this problem.

Damage to the Reservoir Internal Coating.   Care should be taken during the flush, inspection, and manual internal cleaning of carbon steel reservoirs with internal coating. Boots must be removed during internal inspection. Additionally, welding on the reservoir perimeter severely damages the coating.

Addition of Overhead Rundown Tanks.   After the reservoir is installed, an overhead rundown tank may be installed to improve system reliability during power failures. The rundown capacity must be reviewed.

Oil Contamination.   Check that the vents are not tied to the flare system, which can pressurize during upset conditions and dump condensate into the reservoir.

Pumps

Design Considerations

Good pump selection is the heart of an effective and efficient lubricating oil system. If the pump is of the proper capacity, pressure capability, and operating range, pump-related system problems are few. The sizing of oil pumps cannot be done casually. Several important factors must be considered before the proper selection can be made. To determine flow capabilities, one must first know the system oil flow requirements to each machine and its auxiliaries. This must include the maximum transient control oil flow, parasitic relief valve leakage (if applicable), and other oil consumptions.

The total required pump discharge pressure can be calculated by adding all the following factors together:

Maximum required pressure
Static head
Interconnecting piping loss
Pressure-reducing valve loss
Console piping losses
Transfer valve loss
Cooler loss
Filter loss
TOTAL = required pump discharge pressure

In general, the most widely used pump is the positive-displacement screw-type pump. Single-stage centrifugal pumps are the second most common type of pump used. Typically, they are used for moderate to large flows at lower pressure, either alone or as the primary pumps in a multipump (booster) system.

Positive-Displacement Pumps

Positive-displacement pumps must be capable of delivering the required flow at 10°C (50°F) at the pump relief-valve setting. The relief-valve pressure setting plus 10 percent should be a maximum of 90 percent of the manufacturer's maximum differential pressure rating for the lube oil pump.

For positive-displacement pumps, the effects of relief-valve accumulation, low viscosity, and reduced suction conditions must be considered. The flow margin should consider controllability at design and off design conditions and available frame sizes more than fixed percentages.

Characteristics of Positive-Displacement Pumps

Gear Pumps. Spur-gear pumps are relatively cheap, compact, and simple in design. Where quieter operation is necessary, helical or double-helical patterns may be used. Both types are capable of handling dirty oil and can deliver up to about 0.02 m²/s (300 gal/min).

The gear pump is most viable for the following applications:

- At 1,200–1,800 rpm input shaft speeds
- With oil viscosity ranges of 100–500 SSU
- Where lube oil pump discharge pressures are greater than 150 lb/in²

Screw Pumps. Screw pumps are quiet running, provide pulseless flow, are capable of high suction life, ideal for pumping low-viscosity oils, can operate continuously at high speeds over very long periods, and have low power consumption. In addition, they are adaptable to turbine drive. They are capable of delivering up to and above 0.075 m³/s (1,000 gal/min).

Screw pumps are most viable for the following applications:

- Systems with low noise requirements
- Systems where steel-cast pump casings are required
- Systems where pump input speeds exceed 1,500 rpm
- Systems where high-viscosity fluids are predominant
- Systems where pump discharge pressures are 150 lb/in² or less. (Note that screw pumps are available for higher-pressure applications, but gear pumps tend to be more economical and more efficient in higher-pressure applications.)

Positive-displacement pump selection by API 614 criteria is based on 115 percent of the maximum system usage. This is reasonably conservative and provides an adequate control margin. Significant oversizing of pumps should be avoided for reasons of cost and poor driver efficiency. Depending again on size, selections from 105 to 200 percent of maximum capacity have been used.

Because of the design characteristics of screw-type pumps, the flow rate increases as the oil viscosity increases, as shown in Figure 7.9. Therefore, the pump must be capable of delivering the required design performance at an oil viscosity of 65 SSU (based on light turbine oil, ISO Grade 32, and an approximate oil reservoir temperature of 65.6°C (150°F). Also, the pump must handle a cold start with 10°C (50°F) oil at full relief-valve accumulation pressure. The system controls must maintain controllability throughout the temperature, flow, and pressure ranges.

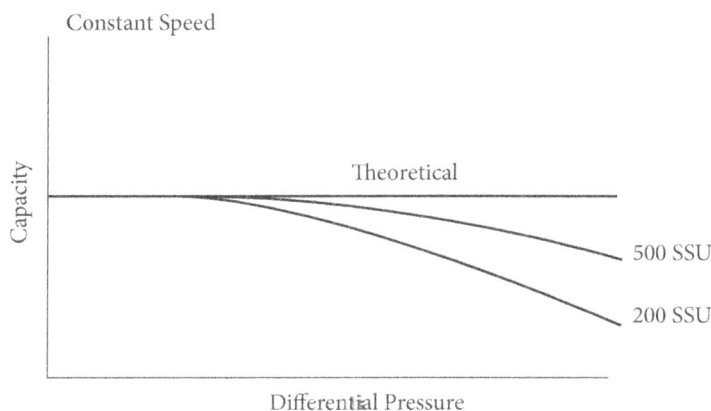

FIGURE 7.9   Positive-displacement pump performance.

Horsepower rating should be sufficient to operate these pumps at relief-valve full accumulation pressure and 10°C (50°F) oil temperature.

Centrifugal Pumps

These pumps have a high rate of delivery at moderate pressures and can operate with greatly restricted output, but protection against overheating is necessary with no-flow conditions. In addition, they will handle dirty oil. The most important design parameters are head variations, startup viscosity, net positive suction head (NPSH) submergence requirements, and pressure and head limitations.

Centrifugal pumps must deliver the specified system pressure across the pump's stable flow range when the oil temperature is 10°C (50°F); the pump must operate at between 50 and 110 percent of its best efficiency point. The pump curve should demonstrate a minimum of 5 percent continuous rise in head from the normal operating point to shutoff.

Drivers are sized, based on API criteria, as follows: horsepower rating should be sufficient to operate the pump at the design system pressure with 10°C (50°F) oil temperature. Centrifugal pump selection must consider the necessary head throughout the console operating range. Care to avoid excessive head is important. Improper application of general user pump specifications can be problematic, resulting in poor system efficiency and higher costs. The pump discharge pressure must be set by determining the maximum required delivery pressure and system losses.

The highest ultimate pressure required in the system must be determined. In many cases, this is the oil pressure required at compressor shutdown and process settling pressure prior to venting. Failure to specify the correct settling pressure or maximum shutdown pressure is a major coordination problem.

TABLE 7.1   System Factors Affecting Choice of Pump Type

Factor	Why Choice Is Affected	Remarks
Rate of flow	Total Pump capacity = maximum equipment requirement + known future increase in demand (if applicable) + 10 to 25% excess capacity to cater for unexpected changes in system demand after long service through wear in bearings, seals and pump. Determines pump size and contributes to determining the driving power	Actual selection may exceed this because of standard pumps available.
Viscosity	Lowest viscosity (highest expected operating temperature) is contributing factor in determining pump size Highest viscosity (lowest expected operating temperature) is contributing factor in determining driving power	May influence decision whether reservoir heating is necessary
Suction conditions	May govern selection of pump type and/or its positioning in system losses in inlet pipe and fittings with highest expected operating oil viscosity + static suction lift (if applicable) not to exceed pump suction capability	For determining positive suction head, or total suction lift
Delivery pressure	Total pressure at pump = pressure at point of application + static head + losses in delivery pipe, fittings, filter, cooler, etc. with maximum equipment oil requirement at normal viscosity Determines physical robustness of pump and contributes in deciding driving power	For determining delivery head
Relief valve pressure rating (positive displacement pumps)	Relief valve sized to pass total flow at pressure 25% above "set pressure." Set pressure = pumping pressure + $70 Kn/m^2$ (10 p.s.i.) for operating range 0-700 Kn/m^2 ($0-100$ lbf/in^2) or, + 10% for operating pressure above 700 kN/m^2 ($100 lbf/in^2$) Determine actual pressure at which full flow passes through selected valve	
Driving power	Maximum absorbed power is determined when considering: Total flow Pressure with total flow through relief valve Highest expected operating oil viscosity Driver size can then be selected	

TABLE 7.2 Pump Performance Factors Affecting Choice of Pump Type

Positive Displacement	Centrifugal
Rate of delivery at given speed substantially unaffected by changes in delivery pressure	Rate of delivery affected by change in delivery pressure
Rate of delivery varies nearly directly with speed.	Therefore flow demand and temperature, which influence pressure, must be accurately controlled
Delivery pressure may be increased within material strength limitation of the pump, by increasing drive power	Because wide variation in output results from change of pressure, pump is well suited for installations requiring large flows and subject to occasional transient surge conditions, e.g., turbine hydraulic controls
Very high delivery pressures can be produced by pumps designed to reduce internal leakages	

Potential Pump Problem Areas

Priming of Positive-Displacement Pumps with Nonflooded Suctions.   In cases where screw pumps are mounted above the liquid level, it is absolutely essential that the pumps are primed for startup. A soft-seated foot valve must be used on the pump suction line to hold oil in the suction line. Screw pumps will seize almost immediately if the pump is started without the rotors being oil wetted. For initial startups and after maintenance, the pump must be manually primed.

Maximum Oil Temperature in Positive-Displacement Pumps During Flush.   The maximum allowable oil temperature to the pump suction is the lower of either the pump physical limitation or viscosity limitation. Physical limitations include seal elastomers, pump thermal growth, material strength, and bearing considerations. Viscosity limitations are related to the minimum oil film thickness, which varies with pump speed and load. A typical limitation is 71°C (160°F) using oil rated at 150 SSU at 37.8°C (100°F). Pump vendors should be consulted to determine this limit.

Slow-Rolling Turbine Drive Auxiliary Oil Pumps.   Under slow roll, the pump will not generate sufficient pressure to open the discharge check valve or relief valve. The result is that there is internal slipping within the pump, and oil does not circulate. High temperature can occur and damage the pump. A properly sized bleeder orifice may reduce the problem, but the installation of steam traps will effectively reduce the pump start time. Continuous-action steam traps should be installed in the turbine inlet piping and casing drains to minimize the start time as well as to provide turbine protection.

Pump Misalignment During Console Shipment.   All baseplates are flexible to some extent and cannot be relied on to maintain factory alignment. This is further compounded by excessive piping runs on the pump discharge as well as poor rigging practices.

Main Pump to Standby Pump Transients.   Automatic starting of standby oil pumps is the greatest contributor to spurious trips of critical machinery. It is not uncommon for a rotating string to trip on low lube oil or seal oil pressure when the standby oil pump (auxiliary oil pump) starts automatically for any of a number of reasons. Although the equipment does not typically suffer any significant damage, the loss of production, the potential for process upsets, and safety considerations make such an occurrence very serious indeed. A reliable standby oil pump startup is of utmost importance in eliminating such spurious trips of the rotating equipment.

The standby oil pump can be started in two different ways:

1. Automatically on loss of oil pressure or level
2. Manually as required by process, utility, equipment, or main-tenancy demands

Automatic standby oil pump starting necessitates the installation of some type of instrumentation that senses oil system pressure or oil level in a vessel. The integrity and reliability of the instrumentation, control logic, and switch set points directly affect the success of an automatic pump switchover.

Switch set points are logically dictated by the location of the respective switch in the oil system. Pressure instruments used for automatic startup of the standby oil pump are typically located in any of three areas:

1. The controlled system header downstream of the oil filters (most common)
2. The pump discharge header
3. The machine in the respective delivery line

Pump startup pressure switches are commonly located in the controlled header downstream of the filters because this provides a stable pressure reference and facilitates the establishment of a workable set point. Generally, this is the recommended location for the start standby oil pump switch(es).

Establishing the setpoint of the switch (on falling header pressure) is critical. If the pressure decay is too small, the pump will constantly start as a result of minor oil system pressure fluctuations. By contrast, if the pressure decay is too large, the lag time between the drop in system pressure and the starting of the standby oil pump will result in a momentary pressure drop at the unit, which will invariably result in an unwanted trip.

Generally, a switch setting corresponding to a 15 lb/in^2 decay works well for systems with controlled header pressures below 300 lb/in^2. Above 300 lb/in^2, it is advantageous to use a decay that is 1–5 percent of the controlled header pressure.

Placement of the standby oil pump start switch in the pump discharge piping presents a new set of problems because pump discharge pressure floats, with oil system resistance dictated by the cleanliness of the oil filters or flow transients. It may be difficult to assign a general number to this particular configuration because of the number of variables involved. However, a pressure decay of 35 lb/in^2 can be used as a benchmark, and further tuning is done during field testing of startup response. Generally, do not locate the pump start switches at any position where the pressure floats.

The standby oil pump start switch can be placed directly on the unit, either in the lube oil supply line, seal oil supply line(s), or both; set points for such switches are typically the same as the respective low-oil-pressure alarm setting. However, the lag time between the start of the standby oil pump and the trip can be so small that a unit shutdown is almost guaranteed.Locating the standby oil pump start switch(es) on the unit is not recommended.

Transient Dampers

Transient dampers are devices that artificially maintain oil pressure for a short time period (accumulators), dampen the pressure signal (hydraulic snubbers), or delay actuation of unit trip circuitry during a pressure drop (electrical time delays).Accumulators can be placed in the controlled header, lube oil supply line, seal oil supply line, or all three. Generally, however, the accumulator is placed in the controlled header, downstream of a check valve to prevent reverse oil flow during system pressure drops. Sizing of the accumulator is based on the startup characteristics of the standby oil pump driver.

A motor-driven pump will accelerate quickly (depending on frame size) and will thus require a smaller accumulator than required by a slower-accelerating turbine-driven pump. An accumulator should provide oil for 3–5 seconds on a motor-drive application and at least 20 seconds on a turbine-drive application. There are many theories about determining the precharge pressure of a bladder-type accumulator. A precharge pressure of 80–85 percent of the start automatic standby pump switch set pressure generally provides satisfactory response during a pump switchover.

When using accumulators in the controlled header, it is very important to verify that the auxiliary oil pump start switch and sensing line for the backpressure-regulating valve are located upstream of the check valve.Location of the pressure switch downstream of the check valve will ensure that the accumulator discharges before the auxiliary oil pump is called on to start, thus rendering the accumulator ineffective. Location of the backpressure-regulating valve sensing line downstream of the check valve will allow the valve to remain open while the accumulator is discharging, thus dumping oil back into the reservoir, instead of diverting oil to the equipment where it is needed to maintain delivery pressures and thus prevent a trip (Figure 7.10).

FIGURE 7.10   Accumulator installation in a controlled header.

An electrical time delay is placed at the low oil pressure trip switch(es) contact outputs such that actuation of the trip switch on low oil pressure will activate a timer. The timer contacts then activate the equipment trip devices after a predetermined time interval if the oil pressure has not been restored. If, however, oil pressure recovers in a short period of time, as is typical of pump switchovers, the timer contacts will be reset by the resetting contacts of the trip switch before activation, and an unwanted shutdown is avoided. Typical time delay settings are 1–3 seconds, but there have been cases where time delays have been set as high as 15 seconds. The use of time delays includes a risk of equipment damage if the delays are not engineered properly, and this must be understood by the user. If indeed this option is pursued, it is strongly recommended that the turbo machinery manufacturer be consulted before implementation.

Regulating and Control Valves

On the oil system, the backpressure regulator is intended to maintain the header pressure under all operating conditions; this includes automatic switchover of the oil pumps. In this case, the controller or valve must respond quickly to prevent the overall system pressure from dropping to the point that a shutdown occurs. However, if response (or tuning) is too fast, the control will overshoot, and control of the system will be lost (the system will now be unstable). Conversely, a slow controller (or valve) will maintain stability in a system, but the set point could drift beyond acceptable limits (loss of control). It is the job of the instrument technician to tune the controller (or valve) to the point where both control and stability are maintained to an acceptable degree. The point to be made here is that a very fast response of the control system does not mean that the system will be stable. The root cause of poor oil system performance is often the pressure-regulating and -control valves. Poor sizing, application, or response can mean the difference between the oil system absorbing transients versus an unwanted turbo machinery shutdown.

Relief Valves

Design Considerations

Relief valves used on positive-displacement pumps are critical to system performance. This is a unique and demanding application. A relief valve must operate smoothly without chatter and re-seat at the specified blow-down pressure. Hydraulic relief valves meet these requirements and are employed most commonly, even though they are neither tight shutoffs or ASME certifiable.

API 614 Specification Requirements.   Relief valves are employed as positive-displacement pump over protection and overprotection for low-pressure delivery lines (downstream of the pressure-reducing valves) on high-operating-pressure lube oil consoles.

Two basic types of relief valves are available for use on lubrication and seal oil systems (Figure 7.11):

- API 614 dictates that pump relief valves all should be designed in accordance with American Society of Mechanical Engineers (ASME) and specified local codes. This is a tight shutoff valve, typically a safety/relief valve, as used in steam or gas service.
- Constant-leakage hydraulic valves are widely used. The slight continual oil flow allows for smooth lifting of the valve plug. These valves are not ASME certifiable under current regulations.

Where the conventional ASME relief valve is used on consoles, excessive chattering, on many occasions, has been experienced during plug lifting, and seat damage has occurred. A great deal of effort has been expended to solve this problem, and demonstrably successful designs are available. However, in oil systems with relatively low pressure levels, the hydraulic-type relief valve becomes more attractive. Although the seat leakage must be designed for, the plug lift is smooth, and instability during lifting is eliminated. Oil system manufacturers have taken exception to API 614 and use these relief valves in view of their smooth performance. These valves can be set only by their accumulation pressures or by careful observation.

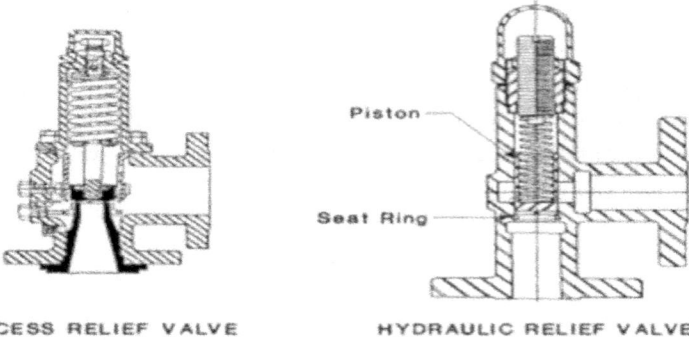

FIGURE 7.11   Process versus hydraulic relief valves.

The addition of a relief-valve bypass valve (usually a globe valve) is very desirable. These bypass valves make pump transfers safer, easier, and more reliable. To accomplish a transfer, the bypass of the idle pump is opened first, and then the pump is started, thus diverting full pump flow back to the reservoir (Figure 7.12). The relief-valve bypass is slowly closed until both pumps are supplying oil into the system, with excess being dumped through the back-pressure regulator to the reservoir. The bypass of the pump that was originally in operation can now be slowly opened, until all the pump flow is being diverted back to the reservoir, and the backpressure-regulating valve is allowed to respond. The pump can now be confidently stopped with no oil system upsets or trips.

Relief-valve bypasses also aid in auxiliary oil pump start-switch testing and horsepower reduction on cold-oil starts. Basically, the bypass provides a means of manually manipulating the pumps for maintenance or testing.

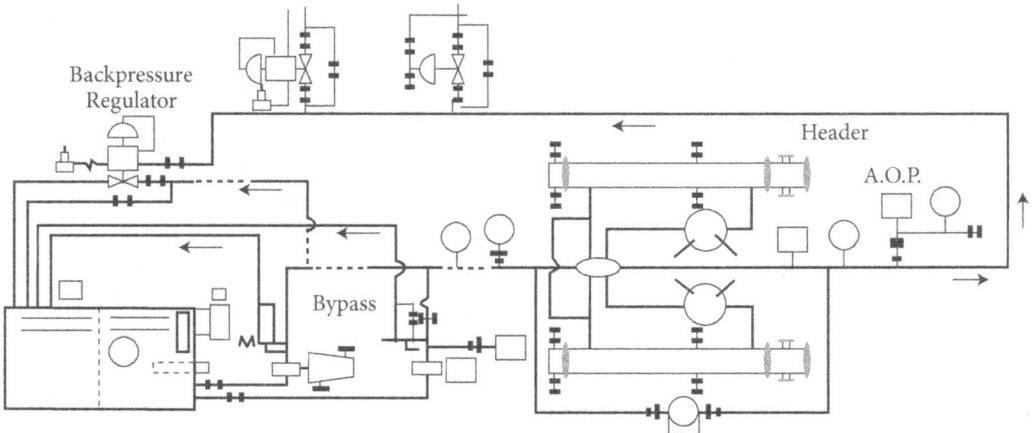

FIGURE 7.12   Relief-valve bypass.

Potential Relief-Valve Problem Areas

Excessive Relief-Valve Leakage Below Relief-Valve Set or Lift-Off Pressure.   Improper field setting of relief valves has caused numerous problems. Improper setting has been determined as the root cause of the following problems:

- Inability of the main oil pump to maintain system pressure
- Inability to switch pump duty without tripping the string on low seal oil to gas differential pressure or low lube oil pressure
- Observance of the backpressure-regulating valve operating fully closed
- Observance of the bearing oil–regulating valve operating fully open (this applies to systems with relief valves installed downstream of the bearing oil regulators, which are used as bearing housing overpressurization protection).

Oil Coolers

Design Considerations

The most important concern in an oil cooler is adequate surface area to reject all the system losses. This is very often overlooked in high-pressure systems. When experience indicates that fouling is not a maintenance problem, many users will employ a single cooler.

API 614 Specification Requirements.   Oil coolers must be designed to remove the total heat added from every source. This includes heat contributed by all bearings, seals, gears, and pumps.

API 614 specifies that shell-and-tube-type coolers used for oil systems should be code constructed in accordance with Tubular Exchanger Manufacturers Association (TEMA) Class C and have removable bundles, steel shells, channels, and covers; tube sheets of naval brass; and tubes of inhibited admiralty. However, these specifications can be changed to suit the need sand preferences of the user. Fixed bundles, U-tubes, different tube pitches, and TEMA Class R have all been used.

The cooler should have vent and drain connections on both the oil and water sides. Oil flows should oppose water inlet flows if system layouts permit. Temperature-control bypass piping may be installed if desired. However, it is important to note that temperature-control valves should fail in such a position as to divert the total oil flow through the coolers.

In designing an oil system, an important item to be taken into consideration is that under normal operating conditions, the oil pressure should be higher than the cooling water pressure. This is obviously to eliminate migration of water into the oil in the event of a tube failure. In lubrication and seal oil systems, where nominal header pressures are in excess of

100 lb/in^2, this is not a problem. However, in straight lubrication systems, where oil pressure requirements are not high, this oil/water factor must be intentionally included.

Temperatures and pressures of the fluid entering and leaving the equipment should be checked regularly. They provide reliable information about the functioning of the unit. For instance, an increase in the pressure drop across the unit (with an accompanying decrease in the temperature range) usually indicates excessive fouling or dirt in the unit. A decrease in the temperature range by itself denotes vapor or gas binding.

Corrosion of heat exchanger parts is of major interest when considering the maintenance of such equipment. Cooling water with a high mineral and solids content will have a high conductivity and a tendency to sludge and form deposition metal surfaces. Corrosion will be seen on parts when the water has a high conductivity. Water containing small amounts of metallic ions, such as iron, copper, and mercury, will cause a pitting attack. Water that is essentially deionizer or free from metallic salts presents no problem.

Means to prevent or retard corrosion typically are water treatments and sacrificial anodes. High or nonuniform corrosion rates leading to tube failures occur at lower pH values. Monitoring of water conditions in the oil coolers and other smaller auxiliary exchangers are often overlooked.

Potential Oil Cooler Problem Areas

Oil heating for flush-past practice of introducing steam directly into the water-side tube bundles to heat the oil is not recommended unless the cooler was specifically designed for this service. Such practice causes differential expansion strains, with possible leakage at the tube joints.

Cooler Thermal Cycling.   Another practice used to reduce the system flush time is to thermally cycle the oil. Most cooler manufacturers recommend not thermally cycling the cooler more than two or three times a day. Manufacturers should be consulted if this procedure is used to flush the system.

Drop in Cooler Efficiency After Cleaning.   Certain coolers, especially smaller units, are designed so that the number of water passes can be varied depending on the bonnet head-position. After maintenance, check for the proper position of the bonnet to ensure adequate heat transfer.

Excessive Tube Failures.   Two main causes of tube failure are fatigue and erosion. Both can be controlled by maintaining the design water flow. Excessive water flow can cause flow-induced vibration, which, if the tube natural frequency is excited, could lead to fatigue failure. Excessive water flow also accelerates tube erosion. A recommendation is to monitor the water flow.

Oil Filters

Design Considerations

Oil filters should be the last component that the oil contacts (except for the pressure-reducing valves) before servicing the intended machinery. For this reason, typical console arrangement provides cooling first and then filtration. This arrangement also allows the removal of any debris from the oil that may have been trapped in the cooler tube bundle.

All API-specified oil systems have twin oil filters. Typically, a filter is piped in series with a cooler, and interconnection with the alternate set is provided by a transfer valve. For bearing design purposes, a minimum oil film thickness of 0. 00075 inch (mil) is typical. Therefore, a filter element capable of removing larger particles is required. The API 614 requirement is 10 μm (0.00039 inch) nominal filtration.

Construction of the filter housing, as with oil coolers, should be ASME code constructed. The maximum design pressure should be in accordance with the relief-valve set pressure or centrifugal oil pump shutoff head.

Oil filters consist of two distinct parts: the housing and the filter cartridges. Housings should be provided with valve vents and drains, a pressure-equalization line, and cover lifts for larger vessels. Filter bypasses should be prohibited. Filter elements should be replaceable and disposable elements, be sized to give a reasonable service life without excessive pressure drop, and changed no less often than every 6 months of service.

The integrity of the oil filters, especially the elements, is an important consideration in ensuring reliable operation of the lube oil system. Neglected or improperly maintained oil filtration systems will result in worn bearings and seals, higher seal oil leakage rates, shortened operating intervals, and major rotor repairs.

Types

Numerous types of filter elements are available today.

Pleated Paper.   Paper (cellulose) filter elements offer the advantage of having very high flow-rate capabilities because of the extended surface area of the pleated-paper design. However, paper elements are sensitive to water and will swell if exposed to water unless treated with a resin.

Pleated Fiberglass/Polyester.   These elements are a crossover application from the hydraulic industry. Fiberglass and polyester elements are not sensitive to water and offer high flow capability without media fiber migration. This is one of the best options for filtration of dirt, sand, and sediments. A polyester pleated-filter cartridge that uses 100 percent synthetic filter media is washable, and thus much of the cost can be saved.

Wound Cartridge Filters. These filters provide true depth filtration, high dirt-holding capacity, and extremely low media migration. They are a superior one-piece cartridge manufactured using a high-speed continuous-wind process and are available in a wide variety of lengths and porosities. Especially significant is the availability of filters in an almost endless combination of media and core material to handle virtually any chemical and/or environment or temperature. The precise winding pattern defines micron ratings and results in higher dirt-holding capacity and efficiency.

Wound filter elements (typically cotton material) offer reliable service. However, cotton fibers are sensitive to water and will swell when exposed to water in the oil, often to the point that the filter collapses. Replacement aftermarket cotton elements are readily available, but the quality of many such elements is very poor.

Ratings

Filtration rating is the single most important criterion to be used when selecting a filter element and is specified in microns. The filtration standard adopted for critical turbo machinery is 10 μm (0.00039 inch). This rating has been selected by the turbo machinery manufacturers based on minimum oil film thickness of journal and thrust bearings. Use of any filter element rated higher than that specified by the turbo machinery manufacturer will compromise the integrity of the unit, even though the element may be less expensive or decrease filter housing pressure drop.

Cartridge efficiency is a measure of how well the element will remove dirt particles greater than the particle size stated and is expressed as a percent. Generally, the higher the efficiency, the better is the element. Three terms are used by the filter manufacturers in specifying cartridge efficiency:

- **Absolute.** Absolute efficiency is the amount of dirt particles at the filter's rating that will be captured and held during one pass. Typical values are 98 or 99 percent. An absolute rating gives the size of the largest particle that will pass through the filter or screen. Essentially, this is the size of the largest opening in the filter, although no standardized test method to determine its value exists.
- **Nominal.** Nominal efficiency is the amount of dirt particles at the filter's rating that will be captured and held during continuous duty or through a closed loop. A nominal rating is an arbitrary size value assigned to a filter by the manufacturer. Tests have shown that particles as large as 200 μm will pass through a nominally rated 10-μm filter. Because the nominal rating is arbitrary, it has no value. Such a rating is more representative of the operation of an oil system. Typical values are 94 or 95 percent.

- **Beta ratio.** The beta ratio is defined as the particle count upstream divided by the particle count downstream at the rated particle size. Using the beta ratio, a 5-μm filter with a beta 10 rating will have on average 10 particles larger than 5μm upstream of the filter for every one 5-μm or greater particle downstream. Beta ratio is the efficiency of an element at a specific particle size but is expressed in whole numbers based on a logarithmic performance curve. For example. A beta ratio of B10 = 50 indicates that an element is 98 percent efficient in removing particles larger than 10 μm. Beta ratios are calculated from actual test data of particle counts upstream and downstream of the element taken during testing.

Though there is no turbo machinery industry standard for testing and evaluating filter elements, there are filtration industry standards for such tests. However, the exact test details and data expressed from such tests may vary between filter manufacturers. A filter element comparison should be based on beta ratio, flow rate, pressure drop, dirt-holding capacity, materials, overall quality, and cost.

The efficiency of a filter can be calculated directly from the beta ratio because the percent capture efficiency is [(beta-1)/beta] × 100. A filter with a beta ratio of 10 at 5 μm is thus said to be 90 percent efficient at removing particles that are 5 μm and larger. Note that the beta rating is meaningless without quoting the size at which it is measured.

A filter's beta ratio does not give any indication about its dirt-holding capacity, the total amount of contaminant that can be trapped by the filter throughout its life, nor does it account for its stability or performance over time. Operating conditions such as flow surges and changes in temperature are also not accounted for; nevertheless, beta ratios are an effective way of gauging the expected performance of a filter.

According to ISO 16889: 1999, absolute filters must have $\beta x \geq 75$ (where x = specified micron size), which is 98.7 percent efficiency. Anything below $\beta x \geq 75$ is considered a nominally rated filter. Such filters can drop as low as a beta ratio of 2, or 50 percent efficiency. As they get more and more clogged, nominal filters eventually start behaving like absolute filters. Nevertheless, inefficient filtration allows hydraulic oil to become lapping fluid that introduces further contamination to a hydraulic system and should be avoided.

Sizing

Correct filter sizing will ensure proper filtration and prevent unnecessary collapsing of the filter elements. This is the responsibility of the oil system manufacturer. However, sizing is ultimately a very important consideration for the user because different filter elements can have significantly different recommended maximum flow rates. If this value is exceeded, the possibilities for problems increase. Collapsing of the filter elements while the unit is online can result in a spurious trip as a result of a rapid drop in oil pressure downstream of the filter housing. Under certain conditions, bearing and seal damage could be experienced. Therefore, it is imperative that the chances of a filter collapse be minimized or eliminated. All reputable filter manufacturers will provide collapse ratings of their filter elements, in addition to the efficiency ratings previously discussed. Numerous factors influence the collapse potential of an element, including:

- **Flow.** Verify that the actual oil flow (per element) does not exceed the manufacturer's rating. For example, a cotton element might be rated at 4.0 gal/min per cartridge, whereas a paper element may be rated at 10 gal/min per cartridge. Obviously, problems will arise if cotton elements are installed in a filter housing that was designed for paper elements. Be especially cautious on systems where the backpressure-regulating valve was not sized for two-pump capacity.
- **Viscosity.** The pressure drop through an element with cold oil will be greater than that at design operating temperature. If the oil is too cold, the filter can collapse.
- **Water.** As mentioned previously, water can cause certain types of filter media to swell, which can lead to collapse.
- **Air.** The presence of air in the filter housing can cause filter collapse from liquid slugging.
- **Pressure surges.** Surging of the oil system, either from instability or from the starting of a second pump, can cause a momentary pressure rise and lead to element collapse.
- **Direction of flow.** In almost all cases, direction of flow is outside the element to inside. If flow is reversed for any reason, the element(s) will essentially explode at a pressure drop that is lower than the collapse rating. In most cases, this will send filter debris into the unit bearings, seals, and control housings.

The best method of controlling filter collapse is elimination of its causes. If water contamination of the oil is a problem, use only a filter media that is not sensitive to water.

Always heat the oil charge to 21°C (70°F) minimum before starting the oil pumps. If this is not practical, open the pump or backpressure-regulating valve bypasses, and recirculate the oil before slowly allowing the oil into the system and through the filters. Do not force cold oil through the filter housings if the differential pressure approaches the collapse rating; throttle the oil until warm enough to continue safely. Always bleed the filters and coolers during startup and prior to a filter switchover.

Probably the most widely used filter element is the cotton cartridge with a cotton-wound matrix and metal core. This type of element provides excellent filtration and can handle small amounts of water. Collapse pressure ratings are usually 90 lb/in^2.

A second type of filter that is commercially available is the acrylic-fiber element with resin binders. These elements are basically impervious to water. However, they have been known to fracture and break during cold-oil starts. Therefore, added caution is necessary when this type filter medium isused. The acrylic cartridge provides excellent service where steam turbine drivers are employed because of its water-resisting qualities (it will not swell). Fiberglass and polypropylene fiber-wound elements are occasionally used as well. Wool should not be used. It has been found in the past that oil flow through wool elements causes a buildup of static electricity. In hazardous environments, this charge has resulted in explosions. Wool is therefore unacceptable as a filtering agent.

In the past, pleated-paper elements were not recommended because of their low water tolerance level. Even the smallest amounts of water caused extreme swelling, thus restricting the oil flow and resulting in high differential pressures to the point of element collapse. The modern resin-impregnated pleated-paper cartridge has greatly improved the water-handling capabilities of pleated-paper cartridges, leading to increasing application in lubrication and seal oil systems. One of the attractive features of the pleated-paper cartridge is its high collapse rating, typically 100 lb/in^2. However, it must be emphasized that only resin-impregnated paper filter elements from reputable sources should be considered in oil systems.

The clean oil pressure drop across a filter is a function of the flow and the oil viscosity. API 614 states that the differential pressure across the filter at 37.8°C (100°F) shall not exceed 5 lb/in^2. This reading is across the filter housing and its elements only. Pressure losses due to piping, coolers, or transfer valves should not be included. Once a filter selection has been made, it is advised that the sizing be verified to check the unit's ability to pass the maximum oil flow at 650 SSU viscosity without reaching the cartridge-collapse differential pressure. This is considered the cold-startup condition.

Maximum dirty pressure drop varies. However, a general rule of thumb is that elements should be changed when the differential pressure rises 15 lb/in^2 above the actual clean-cartridge pressure. Typically, filter flow is from the outside of the element to the inside. There have been instances where piping configuration or transfer valve orientation has resulted in a flow reversal, causing element rupture, system contamination, and confusing differential-pressure indicator readings.

Potential Oil Filter Problem Areas

High Filter Pressure Differential.   API dictates the clean-filter pressure drop to be 5 lb/in^2 at 37.8°C (100°F) and normal flow. The first area to inspect is the sensing points of the differential pressure indicator. A typical installation includes the cooler and transfer valve in the sensing system. For this arrangement, the clean-filter pressure drop would be 20 lb/in² differential.

Cooler	10 lb/in² differential
Transfer valve	5 lb/in² differential
Filter	5 lb/in² differential

The next area that affects the filter differential pressure is the oil viscosity. The higher the viscosity, the higher is the clean-filter differential. Oil viscosity is directly affected first by the actual installed lubricant and second by the oil operating temperature. As presented, the design of the cartridge requires that it not reach the predicted collapse pressure during a cold start.

The final area that directly affects filter pressure differential is the cartridge flow rate. Typically during flush, orifices are removed to enhance flow. Any increase in flow directly affects the filter differential.

Cartridge Separation or Unwinding.   The flow through the filter housing should be in the direction that would collapse the cartridge. Certain transfer valves are designed so that they can be installed backwards. The direction of oil flow through the housing should be checked.

Cartridge Life.   Given the effects of flow erosion, filter-cartridge vendors predict a cartridge efficiency life of 6 months.

Cartridge Ratings.   Owing to the absence of test requirements, 10-μm filter efficiency can vary among various cartridge suppliers. This has led to high filter differential-pressure problems, especially when original an equipment manufacturer (OEM)–supplied elements are replaced with another supplier's elements.

Cartridge Swelling.   This has occurred on both cotton and untreated paper elements in the presence of water in the oil. This is more likely to occur on steam turbine applications because of improper operation of the gland condensers. Maintaining water content below 200 parts per million (ppm) has reduced the high-differential-pressure problem. Cartridges made of acrylic fiber with phenol resin binders usually eliminate the problem. Resin-treated paper elements also have proved successful.

Instrumentation

Design Considerations
The importance of the system instrumentation for monitoring operation, alarming abnormal conditions, and emergency shutdowns is obvious. The controls must enable the oil system to

respond to dynamic conditions both internal and external to the system. OEMs have accumulated a great deal of experience in the unique aspects of oil system instrumentation and system response. Proper coordination and meaningful discussions among users, contractors, and OEMs will ensure a functional system.

API 614 Specification Requirements.   A complete array of instrumentation is necessary to provide reliable operation of an oil system, abnormal operation alarms, and automatic shutdowns. Pressure and temperature indicators, liquid-level gauges, and sight flow indicators provide visual confirmation of console performance.

Switches.   Switches provide an electric signal to activate alarms, start an oil pump, or shut down a rotating string. Certain fundamental functions must be monitored in an oil system. The reservoir should have a liquid-level switch that triggers an alarm at minimum oil level. This indicates that oil must be added to the system and that an oil leak may exist.

With electric heaters, it is prudent to interlock heater controls and level switches. The auxiliary oil pump may have a pressure switch that indicates that the pump is operating. A typical set point would be 25 percent of the relief-valve set pressure or 25 percent of the centrifugal-pump shutoff head. The switch should activate on rising pressure.

As discussed previously, oil filters should be changed when the differential pressure has risen 15 lb/in^2 differential above clean. Typical factory alarm settings are as follows:

- 20 lb/in^2 differential across filter only
- 25 lb/in^2 differential across filter/transfer valve
- 35 lb/in^2 differential across filter/cooler/transfer valve

However, a more accurate field setting can be made by observing the clean-element pressure drop at design oil flow and temperature and then adjusting the switch to alarm at 15 lb/in^2 differential above that value.

In the controlled-header arrangement, the start auxiliary oil pump pressure switch should be located downstream of the filters but upstream of any accumulators. A typical set point is 15 lb/in^2 below the controlled-header pressure but may be as high as 5–7 percent of header pressure on high-pressure oil systems. Switch actuation is on falling oil pressure. It should be noted that a locking relay is required in the start auxiliary oil pump circuitry such that when the pump renews the system header pressure, switch actuation, when returning to normal, does not shut down the oil pump drive. The system design should be such that auxiliary oil pump start is to occur as a result of a fault. The manual reset is to ensure operator attention and prevent cycling.

Lubrication lines must have pressure switches that alarm and trip on falling lube oil pressure. Typical settings, assuming a lube oil delivery of 18 lb/in^2 (gauge), would be a 13 lb/in^2

(gauge) alarm and an 11 lb/in² (gauge) trip. Time delays are user options. They should be limited to a few seconds.

Likewise, seal oil delivery lines must have differential-pressure switches that alarm and trip on falling seal oil differential pressure. Seal oil is the high reference; process gas is the low reference. Typical settings, assuming a design differential pressure of 50 lb/in² differential, would be 30 lb/in² differential to alarm and 20 lb/in² differential to trip. Again, time delays are a user option and should be limited to a few seconds.

A temperature switch may be included downstream of the oil coolers to alarm on rising oil temperature. Typical set points would be 7°C (15°F) above the design oil delivery temperature. As with all electrical devices, switches must be suitable for the *National Electric Code* area classification and weather conditions.

Pressure Gauges.   Pressure gauges are necessary to accurately determine console performance, to aid in setting switches and controls, and to troubleshoot any abnormalities. Typically, a pressure gauge should be installed upstream of the coolers to monitor oil pump discharge pressure. Likewise, a gauge should be installed upstream of the filters to monitor the header pressure and provide a reference for start auxiliary oil pump switch test. The preceding two instruments also act as a check on the differential-pressure gauge installed across the oil filters. Lube oil pressure gauges should be provided, as well as a seal oil pressure gauge and a seal oil/gas differential-pressure gauge.

Thermometers.   Thermometers are used to monitor an oil system's cooling efficiency or inefficiency and to diagnose possible equipment problems. Thermometers should be mounted in the oil reservoir, upstream and downstream of the oil coolers, and at each of the oil throw-off lines of the machinery bearings and/or seals. Stainless steel thermowells are recommended in an effort to ease maintenance and replacement.

Sight Flow Indicators.   These should be installed at every oil drain line of the machinery and in the overhead rundown tank bleed return line.

Liquid Level Indicators.   These should be installed on the oil reservoir, direct contact accumulators, seal oil drainers, overhead seal oil tanks, and degassing tanks.

This is the typical instrumentation. In addition, there is the basic instrumentation needed to control the system's operation. Users will require whatever additional instrumentation that is necessary to provide operators with every means of maintaining reliable oil system operation. This preceding is only meant to provide a guide to nominal instrumentation requirements.

Potential Switch Problem Areas

Motor-Driven Auxiliary Oil Pump.   Cycling on-off-on, check that a locking relay is installed in the auxiliary oil pump start circuit. This circuit should operate in such a manner that once the pump is energized, it must be reset manually.

Hand-Off Auto Switches.   These are used for motor-driven auxiliary oil pumps left in the "off" position.

Unit Trips Before Alarm.   The alarm test orifice required by the API should be checked for integrity.

Switch Failure.   Specifying an explosion-proof switch may permit the design not to be waterproof. Additionally, numerous failures have been traced to improper conduit sealing.

Interconnecting Piping

Interconnecting piping is probably one of the most overlooked aspects of oil system design and installation. However, this area deserves critical attention because improperly designed piping can lead to oil system performance deficiencies and is almost impossible to change after the unit has been installed and commissioned. For this reason, interconnecting piping should be discussed among purchasers, engineering contractors, and vendors during the design phases of a project. However, what should be investigated on the unit that has been installed and commissioned for years, and what can be done if a problem is identified?

The common elements that must be adhered to for an API 614 installation include:

- Compliance with American National Standards Institute (ANSI) B31.3 for pressure design and particularly for proper quality control in materials, fabrication, examination, and testing
- Proper line sizing for design and off-design cases to control the pressure drop within design allowances
- Correct materials considering job-site atmospheric conditions and method and duration of transport and storage
- Adequate support, flexibility, and piping layout
- Proper slope on all lines that may drain by gravity
- Stainless steel pipe downstream of the oil filters (required by API 614) (The standard prohibits socket-weld piping in oil supply lines, requires steel valves, and sets other basic requirements.)

Layout

The physical arrangement of the rotating machinery, oil system, and support systems plays a key role in determining the interconnecting piping configuration. Items to consider include:

- **Arrangement.** Will the rotating string be mounted on a mezzanine or at grade? If grade mounted, attention must be given to adequate elevation differences to allow for adequate fall of the drain piping, along with correct location of intermediate support equipment (e.g., sour seal oil drainers and degassing tanks). If mezzanine mounted, oil system design must account for the interconnecting static and dynamic pressure losses, including losses in long supply headers on large multibody strings of equipment. Other items such as personnel access, overhead equipment (e.g., vessels, cranes, lights, sprinklers), and lay-down areas or openings must be properly considered.

- **Piping slope.** As mentioned earlier, proper slope is required to ensure positive drainage of the oil from the unit back to the oil reservoir. Improper drain line slopes will create traps, cause oil leaks at bearing housings and coupling guards, and could potentially cause flooding of bearing housings. Generally, all drain piping should be given a slope of ½ in/ft of piping run. The exception is pressurized drains (where process pressure can drive drain oil uphill to an atmospheric vessel, as in the case of sour seal oil drainers into a degassing vessel). Inadequate or uphill drain slope from sour seal oil drainers have been known to fill compressor casings and empty reservoirs, whereas the compression string is idle for extended periods.

- **Instrument location.** Alarm and trip switches and seal oil differential-pressure transmitters should be located at the unit if at all possible. If such devices are located with the oil system at grade, compensation for pressure resulting from static and dynamic losses must be accounted for in calibration of the instruments. Obviously, the relative elevations of existing installations cannot be changed easily or economically. However, it is a simple matter to relocate pressure switches to the unit during a scheduled shutdown if problems lie in this area. Items such as piping slope must be dealt with and corrected as the existing configuration permits.

All interconnecting oil piping should be arranged such that all welds can be visually inspected and cleaned. Socket-weld fittings should not be used because they tend to accumulate trash and debris, which will ultimately extend the length of the oil flush and reduce unit reliability. It is also good practice to install blind flanges at the ends of long horizontal runs of pipe to facilitate cleaning. Main piping lines should use flanged joints only; use of threaded connections anywhere except at instrumentation connections is unacceptable.

Material

Piping material is another important consideration in the overall design of oil systems and especially the interconnecting piping. Stainless steel oil supply and drain piping will provide the greatest reliability and the least problems, as will be quickly appreciated on a unit that has a large steam turbine driver. When a steam turbine is a part of the rotating string, there is always the potential that excessive steam leakage from the packing cases may cause water contamination of the lube oil. This water will manifest as corrosion on all bare internal carbon steel surfaces that happen to be wetted by the oil. This naturally includes the oil supply and drain piping. Corrosion in the supply piping compromises the integrity of the bearings and seals of a unit; corrosion of the drain piping compromises the integrity of the oil pumps. The extra cost of stainless steel oil piping at the beginning of a project will pay for itself many times over throughout the operating life of a plant.

Pipe Size

The respective oil system connection size should always be considered the minimum required piping size to be used for construction of interconnecting piping. After the line has been laid out by a piping designer, but before construction begins, the pressure drop at design and maximum flow rates should be calculated.

Typical pipe sizing standards are pump suctions at 3–5 ft/s, pressure feeds at 9 lb/in^2 per 100 equivalent feet of pipe and gravity drains no more than one-half full.

Oil drain piping should be sized to operate no more than half full to ensure good drainage despite possible foaming. Horizontal runs should slope continuously toward the reservoir, and the angle of the slope should be a minimum of ½ inch per foot. Tie-ins to the main drain header should be at a 45-degree angle in the direction of slope. The main header should be provided with a suitable vent. Drain piping is exposed to vapors including air and moisture. This piping should be made of stainless steel or coated internally with corrosion- and oil-resistant pain.

In summary, minimize the use of pipe fittings and maximize the pipeline size. It is never a problem to make the pipe size larger than the oil system supply connection; it is always a problem if it is smaller. Never attempt to make the interconnecting piping size smaller than the connection on the oil system (recall that the oil system designer has selected that connection for a certain pressure drop at the design oil flow rate).

Pump Piping

Both inlet and discharge pump piping must be properly configured to ensure proper auto-matic transfer of pump duty without affecting oil pressure at the unit. Inlet piping must be sized to limit oil velocity to 3–4 ft/s to minimize the chances of pump cavitation. Pump inlet piping must be arranged so as to avoid the collection of air under all operating conditions.

Pump intake of air during startup will increase the pressure drop in the controlled sys-tem header and surely cause a drop in oil delivery pressure at the unit, usually below the trip setting. Proper inlet piping design includes minimizing potential air pockets or traps in the external piping and internal reservoir piping and using eccentric reducers instead of concen-tric reducers, as shown in Figure 7.13. As illustrated by this figure, the concentric reducer has the potential to trap air in the upper portion of the pipe. Additionally, minimizing pump suction bends and fittings helps to reduce piping losses and ensure smooth pump inlet fluid flow (Figure 7.14).

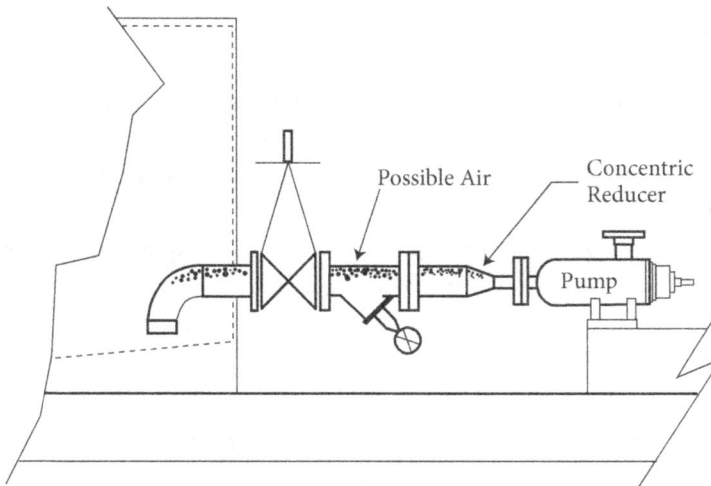

FIGURE 7.13   Concentric reducer in pump suction line.

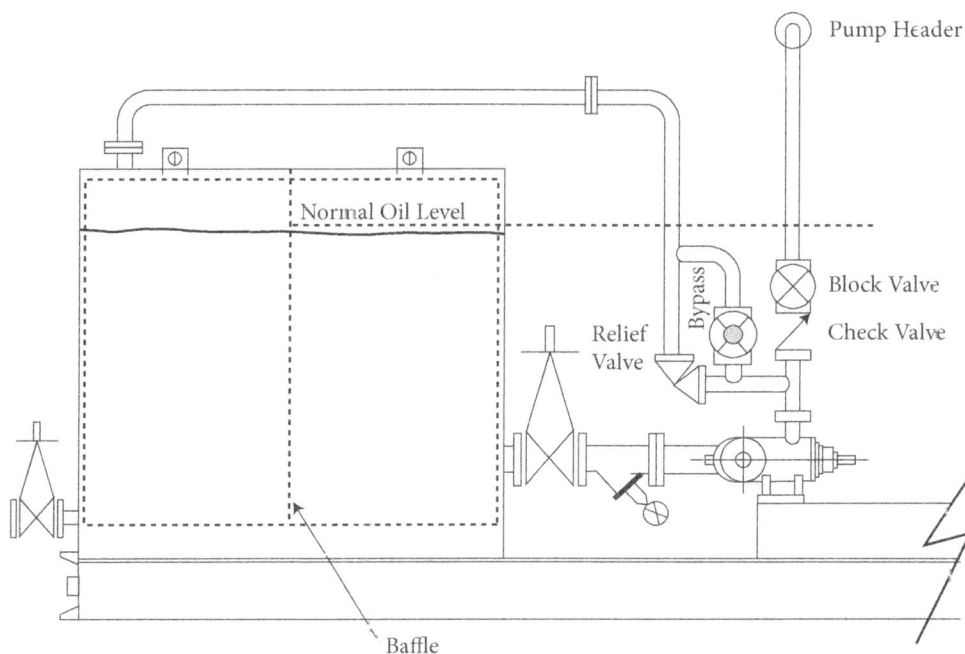

FIGURE 7.14   Locations of pump discharge check and relief valves.

Potential Piping Problem Areas

Drain Pipe.   When a steam turbine is a part of the rotating string, there is always the potential that excessive steam leakage from the packing cases may cause water contamination of the lube oil. This water will manifest as corrosion on all bare internal carbon steel surfaces that happen to be wetted by the oil.

Piping.   All piping used in the system should be made from steel. The use of copper and zinc should be held to minimum because both act as catalysts to promote oxidation and degradation of the oil, although copper alloy tubes are generally used in heat exchangers to minimize water fouling problems and for more efficient heat transfer. The temperatures, normally below 65.6°C (150°F), encountered in turbo machinery equipment do not usually cause copper to have an appreciable catalytic effect on the oil. Minimize the use of pipe fittings and maximize the pipeline size. It is never a problem to make the pipe size larger than the oil system supply connection; it is always a problem if it is smaller.

Piping should be cleaned to remove dirt, mill scale, and weld scale prior to installation and flushing. Cleaning at this time will reduce flushing time in the later stages of the project or overhaul. Generally, new carbon steel piping should be both alkali cleaned (to remove oils, paints, and grease) and acid cleaned (to remove mild rust and scale) and then neutralized and preserved.

Insulation and Heat Tracing.   Insulation and heat tracing are required on oil lines and vessels that do not see a continuous flow of oil. This criterion pertains specifically to overhead seal oil and lube rundown tanks that are exposed to ambient conditions, although it can apply to the entire oil system, depending on the area of installation. Climatic areas that are harsher than described earlier will require insulation and heat tracing of the vessel(s). Heat tracing ensures that the oil will be maintained at a constant temperature approaching the unit design operating temperature. This then guarantees satisfactory operation at design rundown conditions. It is acknowledged that some installations resolve this problem by enclosing the entire unit in a climate-controlled enclosure.

CONCLUSION

Turbo compressors have lubrication systems designed as per API 614 consisting of an oil reservoir, pumps, filter, cooler, and oil lines. Oil is pumped to the lubricated parts (radial and thrust bearings).

Dynamic compressor lubricant should have (1) good oxidative and thermal stability, (2) high viscosity indexes, (3) low pour points for easy cold temperature startup, (4) excellent lubricity for enhanced resistance to friction and wear, (5) extreme pressure lubrication, (6) high resistance to sludge and varnish formation, (7) noncorrosive and stain resistant properties, and (8) compatibility with elastomers and coatings (particularly seal system components, gear unit internal paint, etc). Lubricants with ISO VG 32, 46, and 68 are commonly used in dynamic compressors. Lubrication problems often cause losses in compressor efficiency and increased maintenance costs. These undesirable effects put your compressor's availability and reliability at risk. In the long run, they may harm your processes and reduce your production output.

ICML Questions

1. The difference between minimum operating level and pump suction loss (level at which the oil pump loses prime) in the lube console typically
   a. is 2–6 inches above the pump suction level.
   b. is 2–6 inches below the pump suction level.
   c. is 8–10 inches above the pump suction level.
   d. does not matter.

2. Retention capacity of the lube oil console is
   a. the total volume of oil below the minimum operating level.
   b. the total volume of oil above the minimum operating level.
   c. oil that can be stored in the lube oil console.
   d. 50 percent of the oil that can be stored in the lube oil console.

3. Free surface in the oil reservoir as per API 614 requires a minimum free-surface area of
   a. 0.25 ft²/gal per minute of normal oil flow.
   b. 0.5 ft²/gal per minute of normal oil flow.
   c. 0.75 ft²/gal per minute of normal oil flow.
   d. 1 ft²/gal per minute of normal oil flow.

4. Gravity drain return lines in an oil reservoir must extend
   a. 2 inches below the suction loss level.
   b. 4 inches below the suction loss level.
   c. 6 inches below the suction loss level.
   d. in the same level.

5. To minimize the chances of pump cavitation, inlet pump piping must be sized to have an oil velocity of
   a. 3–4 ft/s.
   b. 10–15 ft/s.
   c. 15–20 ft/s.
   d. 20–25 ft/s.

6. Proper inlet piping design in pump suction requires minimization of potential air pockets or traps in the external piping and internal reservoir piping by using
   a. eccentric reducers.
   b. concentric reducers.
   c. equal tees.
   d. Anything will do

7. All drain piping should be given a slope of
   a. ½ in/ft of piping run.
   b. 1 in/ft of piping run.
   c. 2 in/ft of piping run.
   d. ¾ in/ft of piping run.

8. To prevent reverse oil flow during system pressure drops, generally, the accumulator is placed in the controlled header
   a. downstream of a check valve.
   b. upstream of a check valve.
   c. far away from a check valve.
   d. It does not matter much.

9. An overhead rundown tank may be installed to improve system reliability during
   a. power failures.
   b. normal running.
   c. a choked filter situation.
   d. All of the above

10. To avoid disturbance or turbulence at the pump suction, all oil return lines enter the reservoir
    a. on the opposite side of the baffle from the pump suction lines.
    b. on the same side of the pump suction lines.
    c. very near the suction line.
    d. This does not have any effect.

CHAPTER 8

Gear Lubrication

Gears are used in many applications from high-speed gearing for turbo machinery to slow-speed gear reducers found in agitators. Although gears are used for a wide range of applications, they are perhaps the least maintained of all lubricated components used in process industries. This may result in poor reliability, excessive maintenance and repair costs, and unscheduled production down time. Although gears may be designed by selecting the right material for adequate strength, wear resistance, and scuffing resistance, the choice of lubricant and its application method plays an important role to enhance reliability.

In any gear drive, rolling and sliding types of motion occur simultaneously, but the magnitudes are functions of the type of gear and speed of operation. In an intermeshing gear system, sliding motion leads to the generation of heat. Sometimes, because of inadequate lubrication, the temperature may be sufficient to raise the local temperature to the melting point of the metal. Welding at spots may occur as a result of this high temperature of the tooth surfaces. This may lead to gradual wearing away of the tooth surfaces. To lengthen the service life of the tooth surfaces, lubrication is essential. To prevent metallic contact, the lubricant should have sufficient film strength, and this depends on selection of proper gear material and lubricant.

Basically, the lubricant in a gearbox is intended to serve the following main purposes:

1. To reduce friction and power loss
2. To act as a coolant by dissipating heat
3. To prevent pitting, welding, and breakage
4. To reduce wear of mating surfaces
5. To carry away undesirable contaminants in the effluents
6. To minimize noise, vibration, and shock
7. To prevent corrosion

GEAR TEETH LUBRICATION

In a gear teeth, three types of lubrication conditions are seen: boundary, mixed, and full film. When a gear set starts or stops, boundary lubrication occurs. In boundary lubrication, metal-to-metal contact takes place. If gear sets were operated under conditions of boundary lubrication for prolonged periods, scoring would take place. Hence, the chemical properties of the lubricant are most important to prevent surface scoring.

As relative motion increases, the gearing moves from the boundary to the mixed-lubrication region. Here tooth surface asperities are close enough and have an impact on the coefficient of friction. In this regime, even though wear may occur at a slower rate than in boundary-lubrication conditions, it is still too rapid for a reasonable service life of the gear set.

Optimal lubrication takes place where gear tooth surfaces are completely separated by an elastohydrodynamic (EHD) oil film. Because the viscosity of the lubricant is the most important characteristic in EHD lubrication, selection of the proper lubricant grade is important. Contacting conditions of two gear teeth in mesh, as shown in Figure 8.1, are typical of spur gears.

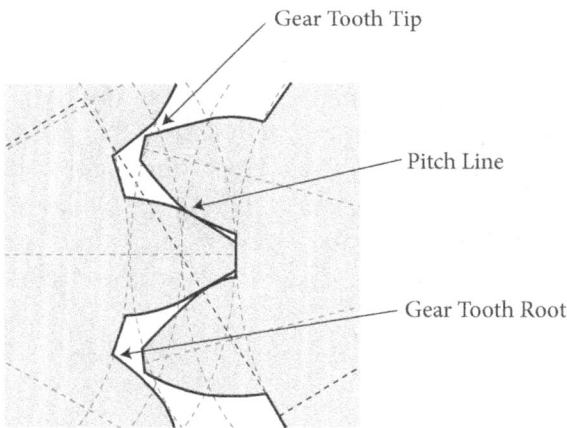

FIGURE 8.1   Contacting conditions of two gear teeth in mesh.

As the contact starts, there is high relative sliding and some rolling. Gradually, sliding decelerates toward the pitch line, and at the pitch line, the motion is only rolling. Again, after the pitch line, sliding takes place and accelerates until the teeth leave the mesh. The radii of curvature of the gear tooth flanks change constantly with the lowest at the root and greatest at the tip.

To evaluate the role of lubricants in the operation of gear sets, it is necessary to understand the contacting conditions of gear teeth in mesh. The tooth surfaces are not perfectly rigid but rather deflect elastically in the contact zone because of the very high contact pressure.

Frictional Force Between the Gear Teeth and the Impact of Speed

In most gears, the frictional force between the gear teeth is typically a combination of sliding and rolling friction. The degree of sliding versus rolling friction depends on the speed of rotation, the applied load, the effectiveness of the lubricant, and how the meshing surfaces engage.

Consider a spur gear with an involute gear tooth profile (see Figure 8.1). Two points of interaction between the meshing gear teeth are important: the tip-to-root contact and the pitch-line contact. In an involute spur gear design, the contact at the pitch line on each gear set is almost rolling friction only. In rolling friction, the two surfaces approach each other in a perpendicular direction (Figure 8.2a). The separation between the surfaces depends on the applied load and the speed in rolling-contact conditions. At higher speeds, the pressure on the lubricant increases, causing a rapid increase in viscosity. As the pressure increases, the lubricant can undergo an instantaneous phase change from liquid to solid. This can result in an elastic deformation of the mating surfaces; as a result, the load is being dissipated across a larger surface area. This helps the gears to transmit the load without mechanical wear because the surfaces are separated by an oil film referred as *elastohydrodynamic lubrication* (EHL).

EHL becomes effective when a lubricant with sufficient viscosity and a high viscosity pressure coefficient is used. If the lubricant has a low viscosity or a poor viscosity pressure coefficient, it may result in metal-to-metal contact leading to boundary lubrication. At lower speeds, approach of the two gear surfaces is too slow to allow the EHL film to form. Under boundary conditions, the lubrication requires the use of extreme pressure and antifatigue additives to prevent wear from occurring. But at the tip to root of the involute gear tooth, sliding friction is the dominant frictional force. In sliding friction, the surface motion takes place in a parallel direction (Figure 8.2b).

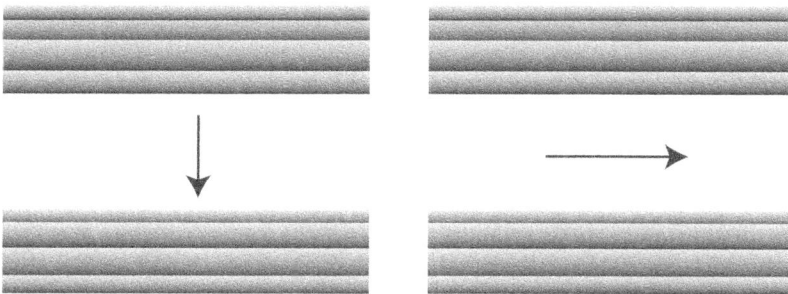

FIGURE 8.2   (a) Rolling and (b) sliding contact between two friction surfaces.

In high-speed gearing, the speed relative to the load is typically high enough that moving surfaces are separated by a full film of oil, resulting in hydrodynamic separation between the moving surfaces. Hydrodynamic lubrication requires the oil to have sufficient viscosity for the applied load and speed, both of which have an impact on oil film thickness. Also, viscosity does not remain constant through the meshing cycle but increases rapidly with pressure.

For low-speed gears, the hydrodynamic lift cannot take place to maintain a full oil film, leading to dominance of boundary lubrication conditions. In the boundary lubrication region, when sliding motion occurs, adhesive wear, galling, or scuffing can takes place. In an involute gear, this typically occurs just above and below the pitch line, and frictional forces transition from pure rolling to sliding friction. Sometimes sliding motion results in high localized temperatures, causing a decrease in the oil's viscosity. This can cause the oil to be wiped away from the converging gear surfaces, which can inhibit the formation of an oil film. This situation leads to boundary lubrication conditions of the meshing gears. Special wear-prevention additives must be used to protect gear teeth if boundary lubrication is anticipated in a gear system.

From a lubrication perspective, it can be concluded that for high-speed gearing, surfaces are separated by a full oil film (hydrodynamic or elastohydrodynamic lubrication). But the slow turning and/or heavily loaded gear drives tend toward boundary lubrication, where point loading can result in surface separation between gear teeth that is equal to or less than the mean surface roughness of the mating gears (boundary lubrication). Table 8.1 provides a general overview of common gear types and the types of lubrication films expected under different loads and speeds.

TABLE 8.1   Typical Lubrication Regimes in Gears

Gear Type	Domination Frictional Force	Low Speed (<100 RPM)		High Speed (>1000 RPM)	
		Pitch Line	Tip/Root	Pitch Line	Tip/Root
Spur	More Rolling ↑ More Sliding ↓	Boundary	Boundary	Elastohydrodynamic	Hydrodynamic
Straight bevel		Boundary	Boundary	Elastohydrodynamic	Hydrodynamic
Herringbone		Boundary	Boundary	Elastohydrodynamic	Hydrodynamic
Helical		Boundary	Boundary	Elastohydrodynamic/ Hydrodynamic	Hydrodynamic
Spiral bevel		Boundary	Boundary	Elastohydrodynamic	Hydrodynamic
Hypoid		Boundary	Boundary	Elastohydrodynamic	Hydrodynamic
Worm		Boundary	Boundary	Elastohydrodynamic	Hydrodynamic

Faulty lubrication leads to many types of the surface defects that may cause heavy functional deterioration. Therefore, correct lubricant must be selected that will help to form a protecting film of lubricant to be maintained uniformly and continuously on the meshing teeth of the gears.

For selecting proper gear lubricant, knowledge of the basic parameters such as type of gearing and operating conditions is essential. Enough viscosity of the lubricating medium is required to develop a suitable oil film between the tooth surfaces and to sustain this film under load. Operating conditions that need to be considered are temperature of the gear teeth, tooth pressure from load, service speed, and external sources of contamination. For proper selection of the type, grade, and method of lubricant application, it is necessary to evaluate the following factors.

- **Type of gearing.** Spur, worm, hypoid, etc.
- **Size.** Pitch diameter, whole depth, and face width as an indication of the surface area to be lubricated.
- **Mounting and enclosure.** Type of housing surrounding the gears. Whether this housing will keep lubricated with splashed oil or oil vapors and keep dirt, water vapor, and other contaminants out must be considered, as well as whether bearings are lubricated by the same system.
- **Speed.** Pitch-line velocity of the gear set.
- **Load characteristics.** Consider whether the transmitted power is continuous, steady, or cyclic. Are shock loads possible? Is high vibration present? Does the unit start and stop frequently?
- **Temperature.** Range over which the lubricant must perform. This would include the lowest ambient temperature anticipated at startup to the highest operating temperature.
- **Environment.** The gearbox operating environment can have significant effects on gearbox reliability.

Gear Design Dictates Lube Design

Gear designs vary depending on the requirements for rotation speed, degree of gear reduction, and torque loading. Transmissions commonly use spur gears, whereas hypoid gear designs are usually employed as the main gearing in differentials. Figure 8.3 shows common gear types. Each gear type has different design advantages, and some have special lubrication requirements.

Spur Gears

Spur gears are normally used in parallel-shaft applications, such as transmissions, because of their low cost and high efficiency. Here gear shafts are parallel, and gear teeth are cut in line with the shaft centerline. In this type of gear, the entire gear tooth makes contact with the tooth face at the same instant. As a result, high shock loading and uneven motion are experienced in this type of gearing. Design limitations include excessive noise and a significant amount of backlash during high-speed operation (Figure 8.3a).

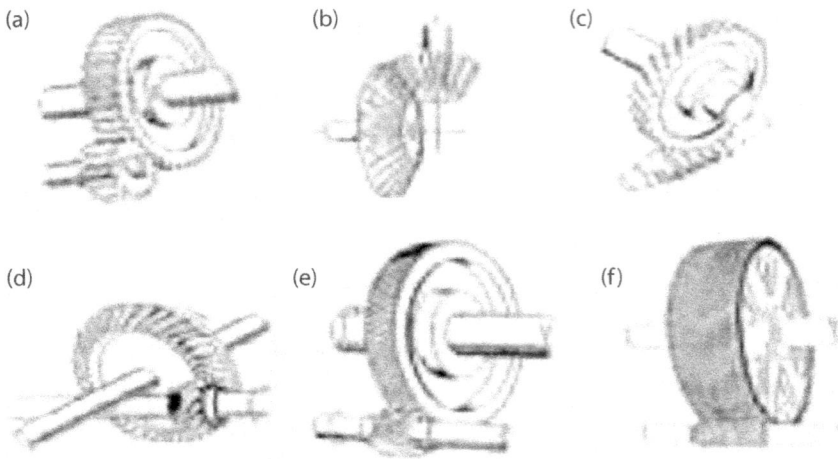

FIGURE 8.3   Common gear types: (a) spur gear; (b) straight bevel gear; (c) worm gear; (d) hypoid gear set; (e) helical gear; (f) herringbone gear.

Bevel Gears

In this type of gear, the motion is transmitted between shafts that are at an angle to each other. These gears offer efficient operation and ease of manufacture and are used mainly for automotive differentials. But because of their noisy operation at high speeds and low load-carrying capacity, they have limited use in industry (Figure 8.3b).

Hypoid Gears

Basically, a hypoid gear is the same as a bevel gear except that shaft centerlines do not intersect (Figure 8.3d). Because of this offset, relative sliding velocity between contacting surfaces is higher than for bevels. Because of this sliding and the high contact stresses, an extreme pressure lubricant compounded with friction modifying additives is required.

Worm Gears

In this type of gearing, the worm resembles a screw thread and drives the worm gear with both elements in the same plane. These types of gears are used primarily to transmit power through nonintersecting shafts and frequently are found in gear reduction boxes because they offer quiet operation and high ratios. Here high-sliding velocity is developed between contacting surfaces on the worm and wheel. Special lubricants containing friction modifiers are used to reduce friction and improve efficiency.

Double Helical Gears

Here gear shafts are parallel, but the gear teeth are cut at an angle to the shaft centerline. The transfer of load from one tooth to the next is more uniform than with spur gears because several teeth are always in contact along some portion of the tooth face at the same time. Because of the helix angle, single helical gear types may generate a thrust load along the axis of the gear shaft. In double helical gears, however, the two helical sections have equal helix angles but opposite hands. Contact conditions are the same as for single helical gears, but because the thrust load from each helix is equal in magnitude and opposite in direction, net thrust load imposed on the gear shaft is zero.

Consider Gear Speed

The speed of the gear box often has an impact on selection of oil viscosity. Normally, high-speed gears require low-viscosity lubricants, and low-speed gears require high-viscosity lubricants. General guidelines on viscosity grades in AGMA-9005-94 specification serve as good rules of thumb. GMA-9005-94 stated

> These guidelines are directly applicable to . . . gears that operate at or below 3,600 revolutions per minute, or a pitch line velocity of not more than 40 meters per second (8,000 feet per minute) . . . and worm gears that operate at or below 2,400 rpm worm speed or 10 meters per second (2,000 feet per minute) sliding velocity.

Gears operating above these speeds are considered to be high-speed gear, and original equipment manufacturer (OEM) recommendations may be sought for lubricant selection. Figure 8.4 is a simple schematic that summarizes how load, speed, and viscosity come together during the lubricant selection process.

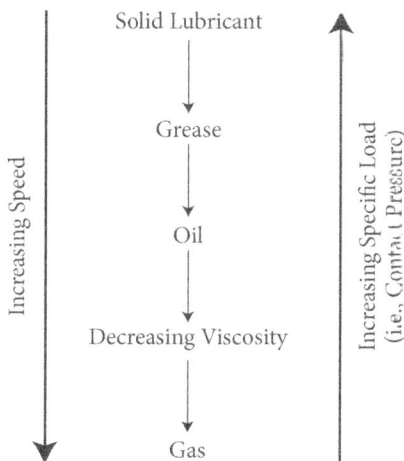

FIGURE 8.4   Effect of speed and load on lubricant selection.

Determine Type of Load

In enclosed gear applications that have small loads, the appropriate oil to be used might be rust- and oxidation-inhibited (R&O) oil only. As the loading increases, the metal-to-metal contact between the gears increases, which may operate in a mixed-film region. For gears operating in mixed-film wear regimes, antiwear (AW) oil is best suited. Again, with increasing load, the gears operate in the boundary regime, and extreme-pressure (EP) oil may be needed for this application. EP oil contains active sulfur and phosphorous compounds that form a protective chemical layer on the gear surfaces when the fluid is compressed out of meshed gears. Care must be taken to see whether there are solid additives in EP oil that may be removed by fine filters and make the oil ineffective.

Identify Viscosity Recommendation

Viscosity is a very important consideration in selecting gear oil. It provides proper thickness of the oil film at operating temperature to keep the mating surfaces of the gears apart during hydrodynamic lubrication conditions. The higher the viscosity, the higher is the load-carrying capability of the lubricant. However, care must be taken in selecting the proper viscosity for a gear application. If the oil is too thick, it will not flow into the gear contact zones. If the oil is too thin, it will be compressed out of the contact zones or fling off the gears while they are in motion. In either case, lubricant starvation will occur, which can result in premature wear-related failures. Also, use of too heavy a viscosity lubricant can result in excessive heat generation, excessive power losses, decreased gearbox efficiency, and improper oil flow. The selection of gear oil also depends on other considerations, such as ambient temperatures, operating temperatures, drive loads and operating speeds.

The primary means of gear lubricant selection is to go with the OEM recommendations. If an OEM recommendation is not available, there are other methods to obtain viscosity recommendations. The first is to use the viscosity ranges recommended by the American Gear Manufacturer's Association in its 9005-E02 standard.

If there is no oil cooler on an industrial gear drive, it is best to determine the maximum expected ambient temperature during operation and (1) increase one ISO viscosity grade if the ambient temperature exceeds 35°C (95°F) and (2) increase two ISO viscosity grades if the ambient temperature exceeds 50°C (122°F). If there is an oil cooler, the maximum ambient temperature is less important because the oil's temperature can be controlled. Therefore, the oil's temperature should determine the viscosity. Therefore, increase one ISO viscosity grade if the oil temperature exceeds 65°C (150°F), and increase two ISO viscosity grades if the oil temperature exceeds 85°C (185°F). If the oil temperature exceeds 90°C (194°F), use a cooler such as a fan or a heat exchanger. The pour point of the gear lubricant should be at least 5°C (9°F) below the minimum expected ambient temperature during startup. A gear lubricant with a lower pour point should be used; otherwise, a heater may be used to heat the oil before starting.

Know Gearbox Construction and Capacity

A gearbox contains various components: the case, gears, bearings, shafts, and seals. The construction of a gearbox, including its geometric configuration, also has an impact on the lubricant selection process. This includes the metallurgies, gear geometries, and cuts of the gears (rough or smooth). Table 8.2 illustrates the part that gear geometry plays in the lubricant selection process.

TABLE 8.2   Effect of Gear Type on Lubricant Chemistry Selection

Lubricant Chemistry	Gear Geometry Type				
	Spur	Helical	Worm	Bevel	Hypoid
R&O inhibited	Normal loads	Normal loads	Light loads and slow speeds only	Normal loads	Not recommended
EP gear lube	Heavy or shock loading	Heavy or shock loading	Satisfactory for use in most applications	Heavy or shock loading	Specified for most applications
Compounded	Not normally used	Not normally use	Preferred by most OEMs	Not normally used	Lightly loaded applications
Synthetic	Heavy or shock loading and/or extreme temps	Heavy or shock loading and/or extreme temps	Heavy or shock loading and/or extreme temps	Heavy or shock loading and/or extreme temps	Heavy or shock loading and/or extreme temps

The gearbox oil capacity also has an impact on the gear oil selection process. If the oil is very costly and the goal is to extend the interval between lubricant drains, the size of the gearbox case is very important. Sometimes the lubricant additives activate at certain temperatures. In a gearbox with large oil capacity, the oil never gets hot enough for the additives to activate, leading to elevated wear. By contrast, in a small gearbox, the oil gets hot quickly, leading to gear additive prematurely activating, oxidizing, and leaving behind deposits. Hence, selecting the right size of the gear case helps in extending the life of the lubricant.

Minimize Effects of the Operating Environment

The gearbox operating environment can have significant effects on gearbox reliability. The operating environment may be hot, cold, dusty, wet, or various combinations of these conditions, all of which may affect oil degradation.

For low-temperatures applications, oil should have a pour point that is 5°C (9°F) below the startup temperature. At higher temperatures, the EP gear oils might result in heavy deposit formation in the gearbox during operation. Water and particulate contamination sometimes finds its way into the gear oil. To minimize these effects, we may take such precautions as using air breathers, sight glasses, and filtration devices. Sometimes the lubricant itself may be selected to compensate for some of the challenges caused by operating conditions.

Ensure Fluid Durability for Extended Drains

Gear oil durability plays an important role in the selection process. The less time a lubricant lasts during service, the more maintenance cost is incurred to replace it. With this hidden cost in mind, users are choosing a gear oil that extends drain intervals. Filtration tools can overcome some of the issues caused by the operating environment to extend the drain interval. In selecting the right oil for your gears, parameters such performance, speed, environmental influences, and special operating conditions have to be taken into consideration.

PROPERTIES OF GEAR OILS

Gear oil properties are determined by the base oil and the additives. The essential requirements for gear oils are described by the leading gear manufacturers and include:

- Operating temperature range
- Viscosity
- Aging behavior
- Low-temperature behavior
- Corrosion protection on steel and nonferrous metals
- Foaming behavior
- Elastomer compatibility
- Compatibility with interior coatings
- Wear protection
- Fretting, micropitting

Operating Temperature

The oil temperature in industrial gears is between 20°C (68° F)and 150°C (302°F) depending on the type of gear and the application. Heating of a gear system, including the bearings and the lubricant, is one of the most important criteria that have an impact on gear performance. In addition, the oil temperature in a gearbox depends on operating conditions. Oil temperatures rise with an increasing ambient temperature and when the oil is exposed to thermal radiation. Temperatures do not get very high when the gear is operated under partial-load conditions or intermittently. For viscosity selection, the oil sump temperature or temperature of the injected oil is an important factor.

Operating temperatures above allowable limits often indicate malfunctions or incipient damage. It is important to ensure that the permissible temperature limits are not exceeded in individual gear components, the lubricant, and accessories (filters, pumps, etc.). When using mineral oil–based lubricants, do not exceed a temperature of 75°C (167°F) to 80°C (176°F).

Viscosity

Viscosity is of primary importance when selecting gear oils because it significantly determines the formation of a lubricant film. Increasing viscosity results in thicker lubricant films, thus improving the antiwear and damping properties, as well as scuffing load capacity.

Viscosity decreases with increasing temperature and rises with increasing load. If the viscosity is too high, increased churning and squeezing losses can result in excessive heat, especially at elevated peripheral speeds. If the viscosity is too low, mixed friction conditions prevail and will result in increased wear.

Viscosity is highly influenced by temperature. The change in viscosity with temperature is usually determined by means of the viscosity index (VI). The higher the VI of a gear oil, the less viscosity changes with temperature, that is, the flatter is the viscosity-temperature (VT) curve. The degree to which viscosity changes with temperature depends on the base oil type, such as mineral oil, polyalphaolefin (PAO), ester, and polglycol, as well as on the VI-improving additives contained in the lubricant.

Aging Behavior

An oil's chemical structure changes continuously when the oil is subjected to high temperatures, mixed with air, or in contact with metal catalysts such as copper, iron, and others, causing it to age. The speed of the aging process depends primarily on the oil's structure and the amount and duration of heat to which the oil is subjected. In addition, contaminants such as water, rust, or dust contribute to oil aging. By adding special additives, the lubricant manufacturer can retard the aging process effectively.

Oil aging is indicated by a change in viscosity and the formation of acids that enhance corrosion and residues. Residues caused by aging occur in the form of lacquer, sludge, or gum, which may clog oil lines, injection nozzles, and filters.

Aging also has a negative effect on an oil's demulsifying capacity, its foaming behavior, its anticorrosion and wear protection, and, to a certain extent, its air-shedding capacity. The aging behavior of oils is determined according to ASTM D 2893.

Low-Temperature Behavior

Depending on the base oil type, lubricating oils solidify at low temperatures as their viscosity increases or as a result of wax crystallization of the contained paraffins. An oil's pour point is indicative of its cold-flow behavior is determined according to ISO 3016. The pour point is the lowest temperature at which the oil still flows when it is cooled down under specified test conditions. In order to ensure rapid and sufficient lubricant supply during a cold start, the lowest temperature occurring in a gear (starting temperature) should always be several degrees above the pour point.

Synthetic gear oils show a much better cold flow behavior than mineral oils. Because of their high VI, synthetic oils are less viscous at lower temperatures than mineral oils with the same nominal viscosity. Their pour point is much lower, sometimes even below –50°C (-122°F).

Anticorrosion Properties

Anticorrosion properties of gear oils are assessed individually for:

- Corrosion protection on steel
- Corrosion protection on copper (compatibility with nonferrous metals)

Corrosion Protection on Steel

If there is water in the system, from either leakage or condensation, it will combine with ambient oxygen and lead to rust forming on inadequately protected steel surfaces. Corrosion on components or rust particles in the oil are returned to the mesh zone and the bearings, where they have an abrasive effect and promote wear. Rust also affects aging stability and demulsification of gear oils and may result in the formation of sludge.

To enhance their rust-prevention properties, gear oils contain polar rust inhibitors that form a compact and protective water-repelling layer. The gear oil's steel corrosion protection properties are determined according to ISO 7120.

Corrosion Protection on Copper (Compatibility with Nonferrous Metals)

For gear oils containing EP additives, it is vital that they do not have a corrosive effect on nonferrous metals, especially copper or copper alloys such as bronze and brass. The corrosion behavior of gear oils is tested according to ISO 2160 using a copper strip. When using a gear oil for components made of or containing copper, such as brass or bronze, it should pass the copper corrosion test according to ISO 2160.

Compatibility with Interior Coatings

Gear housings made of gray cast iron or steel are usually coated to protect them against corrosion during storage, transport, or extended periods of standstill. The primers commonly used for interior coating are resistant to mineral gear oils up to 100°C (212°F). However, they are not always resistant at higher oil temperatures [>100°C (212°F)] or to synthetic gear oils, especially those based on polyglycol. The coatings may get soft, dissolve, or form blisters and chip off, causing gear malfunctions or damage by clogging oil lines, filters, and deaeration holes. Two-component coatings based on epoxy resins normally are resistant to all oil types, even at high operating temperatures. The paint manufacturer should carry out compatibility tests prior to series application.

Foaming Behavior

Gear oils should be able to separate dispersed air rapidly and prevent the formation of stable surface foam. Foam is generated by air bubbles rising to the surface. The bubbles should burst as quickly as possible to keep foam to a minimum.

Particularly in the case of splash-lubricated gears operating at medium to high peripheral speeds, the oil has a pronounced foaming tendency because of the air constantly introduced. Contaminants such as water, dust, corrosion particles, and aging residues may even increase the foaming tendency. Foaming has a strong negative impact on the lubricant's properties, such as oxidation stability, heat dissipation, and so on. Excessive foaming may cause the foam to be forced out of the breather vent; in force-feed lubrication, there is a danger of foam being drawn into the oil pump, causing noise or damage.

The oil manufacturer can reduce the foaming tendency by adding antifoam substances. However, too high a concentration may affect the air-shedding capacity. The foaming tendency of a lubricating oil is determined according to ISO 6247 or ASTM D 892.

Elastomer Compatibility

The materials used for radial shaft seals (RSSs) or static seals (e.g., O-ring seals) must not become brittle or softer when exposed to gear oil because otherwise their sealing capacity would be affected. The seals would suffer premature wear, leading to leakage. Cleaning and possibly expensive gear repairs would become necessary. Especially when higher torques lead to higher operating temperatures or when a gearbox is changed from mineral oil to synthetic oil lubrication, compatibility with the seals should be considered. The tests used for verifying the static and dynamic compatibility of gear oils with elastomers are based on ISO 1817 and DIN 3761, respectively.

Wear Protection

The challenging requirements in gear manufacture today include protection against seizure and micropitting, the reduction of wear under high sliding loads, and the protection of rolling bearings against wear and fatigue.

Protection of Gear Teeth

Heavily loaded gears are potentially subject to seizure and pitting because the high loads generate high pressures and temperatures, which, in turn, can lead to tooth damage and premature gear failure. The risk is particularly high with less than perfect tooth contours and surfaces, impact loads, vibration, a high degree of sliding friction, and high surface pressure.

Scuffing.   The scuffing test according to ISO 14635-1 is generally undertaken to test the capability of gear oils to protect against scuffing damage.

Micropitting.   The micropitting test according to FVA 54/7 has become the industry standard for assessing a gear oil's micropitting load-carrying capacity.

With these interesting conditions in mind, the formulator has numerous ingredients available to build a gear lubricant that provides proper protection. Table 8.3 provides a list of common ingredients used in various combinations in gear oil formulas. A formula does not have to contain every one of these additives, and multiple additives available to provide specific functions. It is the job of the lubricant formulator to choose the proper ingredients to provide synergistic performance in a given application.

TABLE 8.3   Common Ingredients in Various Combinations of Gear Oil Formulas

Ingredient	Function
Base fluid	Mineral oil or synthetic fluid (PAO, ester, PAG) makes up 50–98 percent of the formula.
Viscosity modifiers	VI improvers or polybutene polymers used to increase the base fluid viscosity; these are becoming very common today to replace high-viscosity bright stocks often used in gear oil formulas.
Rust and oxidation (R&O) inhibitors	Rust inhibitors coat metal surfaces to protect against rusting. Oxidation inhibitors defend the oil against degradation resulting from reactions with oxygen in the air when the lubricant is exposed to elevated temperatures.
Copper deactivators	Gear systems can contain some yellow copper-containing metal elements that can be tarnished by gear oil ingredients. These additives protect the metal.
Antiwear (AW) additives	Some gear applications operate in the mixed-film lubrication regime, meaning slight metal-to-metal contact. The base fluid is not sufficient to protect the surfaces from wear, so additives are included to form a sacrificial surface that decreases the friction between the two metal surfaces.
Extreme pressure (EP)	Heavily loaded gear applications can operate in the boundary-wear regime, meaning that the oil film is squeezed out completely. EP additives put down an aggressive coating that carries the load and protects the metal surfaces in lieu of the lost oil film.
Dispersant	These surface-active additives grab onto and disperse contaminants in bulk oil so that they do not collect in the gearbox but instead can be carried to the oil filter for removal. They are used more often in automotive gear oil than in industrial gear oil.
Emulsifiers	These are not common, but some gear oils are formulated to mix with water and stay mixed, such as when water contamination is impossible to avoid.
Demulsifiers	Surface-modifying components are used to promote the separation of water.
Defoamants	Surface-tension-reducing polymeric compounds are used to inhibit the formation of foam on the surface of the oil that could result in housekeeping, oxidation, and wear issues.
Tackifiers	Sticky polymers with high molecular weight are added to increase a gear oil's ability to climb and cling to gears.
Solid lubricants	These are not common in lower-viscosity enclosed gear applications but are often used for lubrication of open gears. EP additives lay down a layer of solids that keep the metal surfaces from rubbing. Common examples include molybdenum disulfide and graphite.
Pour point depressant	Polymeric ingredients are added to modify wax crystals that form in oils at low temperatures. They keep the oil from gelling up, thereby expanding the oil's operability range on the low end.
Compounding additive	A vegetable or animal fatty acid is put into formulations as a friction modifier in formulas for gear applications sensitive to most EP additives.

TYPE OF GEAR OIL

The four types of industrial gear lubricants that could be used in the lubrication of industrial gear drives include:

- Rust and oxidation (R&O)–inhibiting oils
- Extreme pressure (EP) oils
- Compounded (COMP) oils
- Synthetic oils

Rust and Oxidation (R&O)–Inhibiting Oils

The type of lubricant used depends on the type of lubrication regime (hydrodynamic, elasto-hydrodynamic, boundary, etc.) and the type of gear set. For higher-speed applications, where full-film conditions exist, simple R&O-inhibiting oils are used. Aside from their lubricating properties, these oils need to exhibit good oxidation resistance to counter the effects of the heat generated and good corrosion resistance to counteract the effects of any ambient moisture intake, as well as protect any yellow metals that may be present.

R&O-inhibiting oils are commonly used to lubricate high-speed single helical herringbone reduction-gear sets that have pitch-line velocities greater than 17.5 m/s (3,500 ft/min) and are subjected to light to moderate loads. They are also used in the lubrication of spur, straight-bevel, and spiral-bevel gear drives that are subjected to light loads. R&O-inhibiting industrial gear lubricants are ideal for lubricating bearings if both the gears and bearings are lubricated from the same system. Constant relubrication of the gear teeth by the use of either splash lubrication or circulating lubrication systems is preferred because R&O-inhibiting industrial gear oils do not adhere to the surfaces of the gear teeth. They can be used effectively to cool the gear mesh and flush the tooth surfaces of wear particles or debris. R&O-inhibiting gear oils can be easily conditioned with filters and heat exchangers for consistent temperature and cleanliness.

Extreme Pressure (EP) Oils

For slower-speed or higher-loaded gears where full-film separation is simply not possible, EP gear oils should be used. EP additives reduce the coefficient of friction under boundary lubricating conditions.

EP gear lubricants are recommended for use with spur, straight-bevel, spiral-bevel, helical, herringbone, and hypoid gear drives that are subjected to high loading conditions, moderate to high sliding conditions, and high transmitted power conditions. Because some types of EP gear lubricants contain chemically active additives, care must be taken if they are used in systems where the gears and bearings are lubricated from the same system or if they are used in heavily loaded worm-gear drives. EP gear lubricants can contain active chemistries that are corrosive to brass or bronze components. When used in these applications, the lubricant supplier should be contacted to determine if the EP gear lubricant can be used in such applications.

Some EP gear lubricants will also contain solid lubricants such as graphite or molybdenum disulfide that are held in a suspension. These solid lubricants are formulated into the industrial gear lubricant to further improve the gear lubricant's load-carrying capabilities. When EP gear lubricants that contain solid lubricants are used, care must be taken if fine filtration is used. Extremely fine filtration can remove solid lubricants. Ideally, if an EP gear lubricant containing solid lubricants is going to be used, the solid lubricants should be collodially suspended and have a particle size no greater than 0.5 μm.

EP gear lubricants should never be used in industrial gear drives that have internal backstops, such as those found on conveyor belts, or in the lubrication of cooling-tower gear drives that employ ratchets. The EP chemistry will not allow the clutch or sprag mechanisms to properly engage, resulting in slippage. This can cause serious safety hazards, such as a conveyor belt continuing to turn or slip after the enclosed industrial gear drive is shut off. EP gear lubricants can be easily conditioned with filters and heat exchangers for consistent temperature and cleanliness.

Compounded Gear Oils

Compounded gear oils are used primarily to lubricate enclosed worm-gear drives, where the high sliding action of the gear teeth requires a friction-reducing agent to reduce heat and improve efficiency. The surface-active agent, which is a fatty or synthetic fatty oil, prevents sliding wear and provides the lubricity needed to reduce sliding wear. Their use is limited by an upper operating temperature of 82°C (180°F). For worm gears, compounded oil is preferred over an EP oil because chemically active EP additives can be corrosive to yellow metals (brass, bronze, etc.), which are commonly used for the ring gear in worm drives or in bearing cages. Constant relubrication by either splash lubrication or circulating lubrication systems of the gear teeth are preferred. Most worm-gear drives normally require an ISO 460 or 680 compounded oil and, in some cases, an ISO 1000 oil. The viscosity grade required depends on the worm-gear drive's speed and operating temperature. Generally, the lower the worm's gear speed, the heavier is the needed viscosity grade.

Synthetics

In some instances, such as high or low operating temperatures or the desire to extend oil drain intervals, the use of high-performance gear oils (i.e., synthetics) is recommended. Where their usage can be financially justified, these lubricants offer significant benefits in low-temperature startup, high-temperature thermal and oxidative stability, and improvements in film strength. Although there are several different types of synthetic gear oil, the two most common are polyalphaolefin (PAO) and polyalkylene glycol (PAG). Both have their relative merits, but their use should be considered judicially because the cost of switching to a synthetic gear oil often can far outweigh the benefits. When selecting synthetic gear oil, it's not uncommon to drop down one ISO viscosity grade from the OEM recommendation because the effective viscosity of a synthetic gear oil at elevated operating temperatures often matches that of the OEM-recommended mineral oil grade because of the higher VI of synthetic oils.

Synthetic gear lubricants offer the following advantages in enclosed gear drive applications:

- Improved thermal and oxidation stability
- Improved viscosity-temperature characteristics (high VI)
- Very good to excellent low-temperature characteristics
- Lower volatility and evaporation rates
- Reduced flammability (depending on the type of synthetic base used)
- Improved lubricity at mesh temperatures above 185°C (365°F)
- Resistance to the formation of residues and deposits at high temperatures
- Improved efficiency owing to reduced tooth-related friction losses (low traction coefficients)
- Lower gearing losses owing to reduced frictional losses (low traction coefficients)
- Extended oil drain intervals
- Reduced operating temperatures, especially under fully loaded conditions
- Reduced energy consumption

AVOIDING COMMON GEARBOX LUBRICATION PROBLEMS

Lack of lubrication and use of incorrect lubrication are the two top causes of premature failure in gearboxes. With splash lubrication, it is important that a certain oil level is maintained at all times to ensure damage-free operation. If the oil level is too low, it may result in starved lubrication, inadequate heat dissipation, and increased wear. If the oil level is too high, churning losses may increase, resulting in higher oil temperatures and gear friction losses. In order to keep churning losses low, the depth of immersion is reduced with increasing peripheral speed. Further oil lubrication methods are immersion circulation lubrication and force-feed lubrication. In applications where peripheral speeds are too high for immersion lubrication and for most gears running in plain bearings, force-feed lubrication injecting oil directly to the mesh is used. Confirming how much lubrication is required for a given gearbox in a specific mounting orientation is relatively easy. Gearbox manufacturers include this information in their catalogs and manuals.

Other than incorrect lubrication and lack of lubrication, controlling contaminants within gear drives helps to enhance gearbox reliability. Most contamination-control strategies for gear drives focus too much on coarse particles. In this context, *coarse* refers to particles in excess of 20 μm in size. In this size range, even though particles are invisible to the naked eye and are, on average, three to four times smaller than the cross section of a human hair, damage can still occur

The hygroscopic nature of oil makes it next to impossible to keep gear oil completely free from water, keeping water at or below the saturation point is key. For many conventional gear oils, the saturation point of the oil at typical gearbox operating temperatures ranges around 400–600 parts per million (ppm) of water (0.04–0.06 percent by volume). For a gearbox that holds approximately 20 liters (5 gal) of oil, which equates to as little as 1½ teaspoons of water. Once the saturation point is exceeded, water will come out of solution into either the free or emulsified state. In this condition, the deleterious effects of water, which include loss of film strength, rust, and corrosion, increase exponentially, seriously impacting equipment life.

This problem is most pronounced in gear drives that operate intermittently at low ambient operating temperatures. Whereas 500 ppm of water in a gearbox operating at 60°C (140°F) typically will be all in the dissolved phase, shutting down the gearbox and allowing the oil to cool to 0°C (32°F) will cause most of the water to come out of solution.

Water also has a secondary effect on gear oils. Many of the additives used in gear oils are either water soluble or react with water. As such, whenever gear oil is left saturated with moisture owing either an extended shutdown period or inappropriate new oil storage, additives can be either stripped or rendered ineffective. For most gear oils, water levels need to be kept dry enough that any water that may be present does not form emulsions or free water. While not possible in some circumstances, practical limits for water in gear oils should be below 200–300 ppm (0.02–0.03 percent by volume).

GEARBOX CONTAMINATION CONTROL

Controlling contaminants within gear drives requires a concerted effort to assess each possible source. Even something as simple as changing the oil can result in a significant amount of particle and moisture ingress unless the utmost care is taken. The first step in controlling contaminants is to review all possible sources. These include both contaminants introduced from the outside and contaminants created internally. Some of the more common sources include:

- Airborne dirt and moisture
- Water from wash-down/sanitation
- Water from the production process
- Unfiltered new oil
- Internally generated wear debris
- By-products of oil degradation

With any contamination control strategy, the first place to start is to look at external sources. Most external contamination in gearboxes comes from the breather/vent port. This is so because many gearbox designs use a combination breather and fill port. Careful examination of the fill port/breather cap often reveals little more than a course sponge or wire wool/mesh to restrict contamination ingress. Wherever possible, older-style breathers or combination breather/fill ports should be replaced with modern, high-efficiency breathers. However, in most plants and industrial environments, moisture is an issue, especially because many gear oils are hygroscopic. Where airborne humidity or process water ingression is an issue, it is necessary to remove moisture from the air as it enters the gearbox headspace. This requires the use of desiccant breathers, which include both a particle-removal element capable of elim-inating silt-sized particles and a desiccating medium, often comprised of silica gel, to remove all traces of moisture from the air as it enters the gearbox.

Although desiccant breathers are effective for removing particles and moisture from the air, in some environments where a lot of moisture and humidity is present, the life expec-tancy of the silica gel can be a little more than a matter of weeks. Under these circumstances, a more cost-effective solution may be the use of a hybrid breather that remains sealed when no air exchange is required. In this case, thermal expansion and contraction of the headspace as the gearbox heats up or cools down are controlled via a bladder that expands or contracts to equalize pressure. If a significant pressure differential exists, for example, during startup, a series of check valves on the bottom of the breather open to equalize pressure between the gearbox headspace and the environment. Unlike standard desiccating breathers, the advan-tage of hybrid breathers is that the system is nominally sealed, preventing contamination ingress and preserving the life of the breather. Depending on application and environment, these so-called hybrid breathers can last as much as 5–10 times longer than a conventional desiccant breather.

Having a desiccant or hybrid breather and removing other sources of contamination are excellent first steps in any gearbox contamination-control strategy. Eventually, however, there will be a need to open the gearbox to change the oil, check the oil level, and so on, and in so doing, it is easy to undo all the benefits provided by high-quality breathers. To illustrate this point, consider the way oil is changed on most splash-lubricated gearboxes.

Contamination During Oil Change

Because the oil must be changed with the gearbox shut down, the oil inside the gearbox is typically colder than during normal operation. As the oil cools, the viscosity increases, making it difficult to drain all the old oil out of the drain port. To reduce the amount of time it takes for the oil to be drained, most mechanics are prone to removing the breather or fill port to increase flow rate. However, by doing so, the effect on contamination control can be disastrous. Draining 5 gallons of oil from a gearbox requires the equivalent volume of air to enter through the open port, which in most ambient plant environments is enough to increase the effective ISO cleanliness code and moisture content within the gearbox by several orders of magnitude.

Gearbox Maintainability

The solution for controlling contaminants is to configure the gearbox to remain sealed during all phases of normal operation, including routine planned maintenance such as level checks and oil changes. This can be easily achieved by modifying the drain and fill/breather ports with simple adapter kits that permit multiple access points to the gearbox without opening the gearbox sump to the environment. This kit, which is used to replace the breather/fill port, allows for the installation of a desiccant breather and quick-connect fittings to facilitate the addition of new oil without opening the gearbox. By combining this adapter with a simple quick-connect fitted on the drain, this gearbox can be maintained without ever being exposed to the ambient environment.

Maintaining the correct oil level is also critically important, particularly in smaller, splash-lubricated gearboxes where an oil level variation as small as ½-inch (12 mm) can mean the difference between success or lubricant starvation. Because of this, routine oil level checks are an important part of any gearbox preventive-maintenance program. To facilitate level checks, many gearboxes are equipped with a dipstick level indicator. High and low (shutdown and running) levels should be marked on any level gauge to indicate the correct oil level under any operating condition.

For gear drives, the secret for precision contamination control is the use of supplemental offline filtration. This simple strategy, which involves taking a small amount of oil from the wet sump, passing the oil through a high-efficiency filter, and returning it back to the sump, has proven to be very effective at maintaining optimal levels of cleanliness in gearboxes.

The simplest solution is to use a permanently installed bypass filtration system. This system has a pump and two filter housings, the first housing being used to remove either water or large particles and the second filter rated to remove silt-sized particles. Flow rates for this type of system should not exceed 10 percent of the total oil capacity (e.g., no more than 5 gal/min (1.3 liter/min) for a 50-gallon (190-liter) sump), but even with just a small amount of oil passing through the filter at any given time, these systems can effectively control particles and moisture to very low levels.

CONCLUSION

Compared with other predictive maintenance tools, such as vibration analysis and thermography, oil analysis is typically a leading indicator of a wear problem in gear boxes. It often shows any significant change in component or lube condition weeks or months in advance of problems. This is important for low-speed gears and bearings in a multi-reduction-gear drive where vibration analysis is typically less sensitive because of difficulties with slow-speed accelerometers. In many cases, oil analysis will typically show the problem first, but because vibration analysis helps in localizing the exact component and failure mode, both may be used for an effective predictive-maintenance program.

ICML Questions

1. In a gear drive, the motion occurs is
   a. sliding type.
   b. rolling type.
   c. both sliding and rolling.
   d. tangential rubbing.

2. Basically, the lubricant in a gearbox is intended to serve the purposes of
   a. acting as a coolant by dissipating heat.
   b. reducing wear of mating surfaces.
   c. minimizing noise, vibration, and shock.
   d. All of the above

3. A high-speed spur gear pitch line encounters
   a. boundary lubrication.
   b. hydrodynamic lubrication.
   c. elastohydrodynamic lubrication.
   d. All of the above

4. Normally, high-speed gears require
   a. low-viscosity lubricants.
   b. high-viscosity lubricants.
   c. extra-high-viscosity lubricants.
   d. Viscosity has no role to play.

5. For gears operating in mixed-film wear regimes,
   a. high-viscosity oil is best suited.
   b. antiwear (AW) oil is best suited.
   c. oil with a high pour point is needed.
   d. oil with fewer foaming characteristics is needed.

6. For gears with low-temperature applications, oil should have a pour point that is
   a. 5°C (9°F) below the startup temperature.
   b. 5°C (9°F) above the startup temperature.
   c. 10°C (9°F) above the startup temperature.
   d. the same as the startup temperature.

7. Practical limits for water in gear oils should be
   a. below 200–300 ppm.
   b. below 400–500 ppm.
   c. below 600–700 ppm.
   d. below 550–650 ppm.

CHAPTER 9

Storage and Handling of Lubricants

Lubricants are delivered to end users in containers, drums, and barrels and sometimes in bulk. At each stage of storage, handling, and final distribution, good practice can be implemented to enhance the performance and service life of the lubricants. Some methods are necessary for reasons of hygiene, safety, and environmental impact, whereas others ensure that the lubricant is not contaminated when it is about to be used. Plant owners can invest in product storage and handling practices, although this is often overlooked.

The important issues that affect lubricants as a result of improper storage and handling are as follows:

1. Pollutants such as dust and water may enter the lubricant if adequate precautions are not taken. These pollutants then get into the lubricated equipment.
2. If the lubricants are not labeled properly, lubricant identification become difficult. As a result, an unsuitable lubricant may be added to a machine, causing damage to the equipment.
3. Mixing some lubricants together may cause equipment breakage.
4. Damaged packaging can result in leaks that can affect the environment.

It is relatively easy to avoid these potential problems if you take a look at your storage and handling procedures. These procedures can have a significant impact on the success of your lubrication program. By applying some basic rules that are usually a matter of common sense can improve the performance and service life of lubricants. The good practices that plants can adopt to improve storage and handling issues include:

- New lubricant quality control (QC)
- Proper unloading and handling
- Proper storage environment
- Proper storage management

- Accurate labeling
- Health and safety precautions
- Fire precautions
- Contamination control
- Proper long-term storage
- Prefiltering of lubricants
- Use of sealable, cleanable top-up containers
- Precautions during top-up activities
- Appropriate stock rotation

NEW LUBRICANT QUALITY CONTROL

A best practice in lubricant storage and handling begins with verifying the lubricant quality as the lubricants are delivered to the plant. Test methods should be rigorously defined and repeatable, and the instruments should be routinely calibrated to a recognized standard. The key test parameters, the physical properties and contamination levels, could be measured fairly easily with relatively inexpensive on-site instruments. If time permits, it would be better to verify on-site results against those provided by a professional oil analysis laboratory before condemning large shipments.

PROPER UNLOADING AND HANDLING

The best way of unloading lubricants is to use a platform or fork lift truck. Oil or grease drums weigh more than 180 kg (390 lb). Because of their weight and size, they are difficult to handle. The drums must be unloaded by fork lift. If there is no fork lift available, it is strongly advised that wooden or metal ramps should be used to skid the drums gently down to the floor, as shown in Figure 9.1.

FIGURE 9.1   Proper unloading of lubricant drums.

The drums should not be dropped from the bed of the truck onto a pile of tires. A fall like this can damage the welds or the walls or even cause the drum to burst. Once the drums are unloaded, they should not be rolled on the floor for short-distance transportation. In any case, drums, especially grease drums, must not be rolled overlong distances. This would cause damage to the welds, ferrules, and ends of the drums, which could cause leaks to occur. Therefore, it is preferable to use a drum-carrier cart (Figure 9.2).

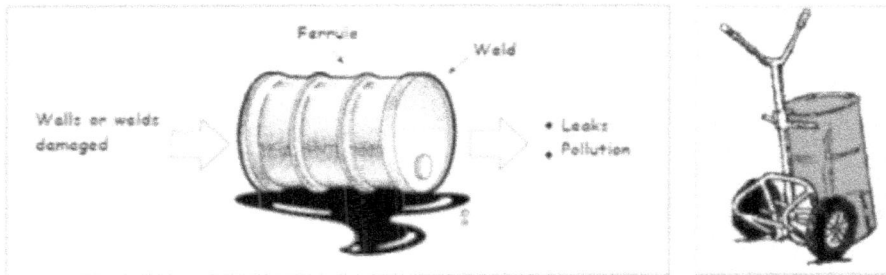

FIGURE 9.2   (a) Rolling drums can be dangerous. (b) Use a drum-carrier cart to move drums.

PROPER STORAGE ENVIRONMENT

Whether your lubricants are stored indoors or outside, environmental conditions can reduce their shelf life. Fluctuating temperatures may cause a reaction known as *thermal siphoning*, in which air moves in and out of the container's headspace and the atmosphere. Moisture and airborne particles travel with the air, resulting in contamination and degradation of the lubricant. The fluids are also vulnerable to contamination by dust and dirt. Extreme hot or cold temperature can lead to chemical degradation (Figure 9.3).

FIGURE 9.3   Lubricants stored indoors, where the environment is better controlled.

All lubricants should be stored in a room or building designed for the purpose. Ideally, lubricants should be stored indoors in a clean and dry environment where conditions can be better controlled. The storage space must be lighted and ventilated sufficiently, if possible, with dust-free air. The location must not be exposed to extreme hot or cold temperatures; moderate temperatures should be maintained.

If outdoor storage cannot be avoided, lubricants should be sheltered as much as possible from environmental conditions that may degrade them. For better outdoor storage, a roof or tarpaulin can help shelter the lubricants from rain, precipitation, or heat generated by direct sunlight. It is also better to place drums or containers on blocks or racks to protect them from ground moisture.

When not in use, it must be ensured that each lubricant package (drum, pail, etc.) is staged for protection from the environment by use of quality breathers and by maintaining closure on all openings (valves, hoses, and pipes). It is advisable to have retention areas designed to prevent any possibility of pollution in case of leakage.

PROPER STORAGE MANAGEMENT

For easier inventory management, lubricants should be stored in a single indoor location. The floor of the storage area must be flat and hard built. It is advisable to use drum-handling equipment to put the drums in the racks. Aisles should be wide enough for forklifts or other equipment used to handle/move lubricant containers. Easy access to all products must be ensured. Proper shelves and racks should be used to hold, display, and protect containers. However, if there is not enough room in the shelter and some products must be kept outdoors, a number of precautions must be applied. When the lubricants are stored outdoors, the stock must be kept to a low limit to encourage rotation to avoid exposure of the products to bad weather and pollutants over time. It is highly advisable to also have fire extinguishers in the lubricant storage area.

Indoor Storage

Whether kept inside (preferable) or outside, drums are best stored on their sides (horizontally) with bungs at the three and nine o'clock positions. This prevents moisture from accumulating on the top (especially outdoors) and also minimizes the ingress of air through the bungs (indoors or out). Place the plugs on a horizontal line (nine and three o'clock), as shown in Figure 9.4.

The plugs must be on a horizontal line to prevent the penetration of air or condensation water into the drums. This also prevents moisture from accumulating on the top (especially if drums are kept outdoors).

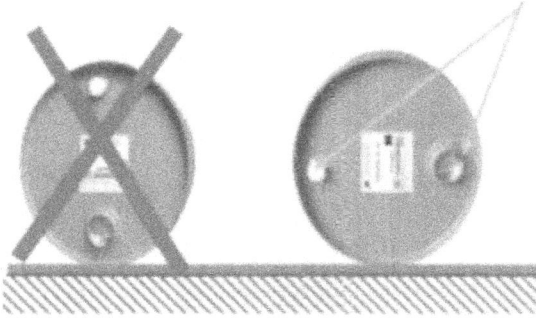

FIGURE 9.4   Proper storage of drums. Plug should be in horizontal direction.

Outdoor Storage

When drums are stored outdoors, it is necessary to store the products on a sealed surface, for instance, a hard-built flat slab, for environmental reasons. This will prevent pollution of the soil and any groundwater in the event of accidental spillage. It is also preferable to use storage racks with protection from bad weather or outdoor storage cabinets.

What Happens with Vertical Storage?

If the drums are stored vertically, rain will accumulate on top of them. Because the sun heats the product in the drum, it will expand and fill part of the airspace inside the drum. Accordingly, at night, when the temperature drops, the volume of the product will decrease in the drum causing a strong suction effect. As a result, the water that has accumulated on top of the drum is drawn in, mixes with the lubricant, and is decanted to the bottom of the drum (Figure 9.5).If drums must be stored upright outside, use drum covers (rain caps) or store them at a tilt to prevent moisture from pooling and entering through the bungs.

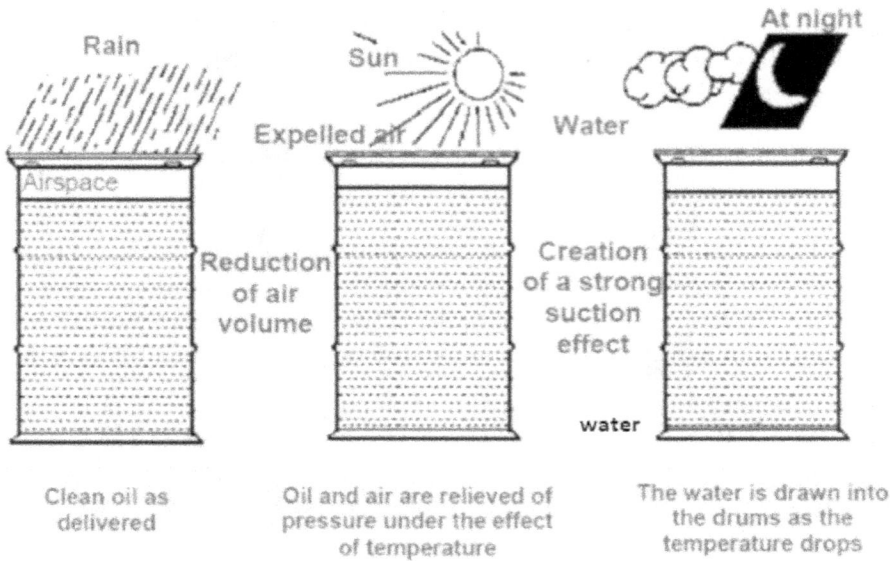

FIGURE 9.5 Vertical outdoor storage.

The drums must be placed on beams at different heights and the plugs turned suitably to prevent any possibility of water getting in. This is particularly the case with grease drums that must be stored vertically (Figure 9.6).

FIGURE 9.6 Proper outdoor storage of grease drums.

Lubricant Degradation During Storage

Lubricants can degrade in storage mainly for the following reasons:

1. Contamination, most frequently dirt and water
2. Exposure to excessively high temperatures
3. Long-term storage

Entrained water promotes base oil degradation and additive depletion. Free water provides a place for microbial contamination that is corrosive and harmful to lubricant properties. Emulsified water also has a tendency to impair air-release properties of oil. When air fails to detrain (release air to the headspace), a common consequence is oil oxidation. Solvents, fuels, and other incompatible lubricants are harmful for lubricating oil.

Thermal Degradation of Lubricants

Most good-quality synthetic and conventional mineral oils are not affected by storage temperatures below 49°C (120°F). However, storing lubricants near furnaces, steam lines, or direct sunlight in high-temperature climates for a prolonged period may cause additives and base oils to oxidize prematurely. A significant darkening of the oil color is an indicator of this condition. In greases, the oil may begin to separate from the thickener; this is known as *bleeding*. The separated oil will typically appear on the surface of the grease, depending on the type of thickener used. Lubricants that are potentially contaminated with volatile products, including diesel fuel, kerosene, or any other solvent, must never be stored in high-temperature environments. The presence of solvents can be identified by a test called the *flash-point test*. In addition to evaporation and fire hazards, they can distort or even burst the storage vessel if tightly sealed.

Lubricants have shelf-life limits that vary from product to product and from manufacturer to manufacturer. The shelf lives of lubricants are affected by the blend process, which has an impact on whether additives are fully dissolved or are susceptible to settling during transport and storage. Contamination from rainwater, large temperature swings, and high-temperature storage methods also shortens shelf life.

ACCURATE LABELING

Are the lubricants clearly labeled and easily distinguishable? Lubricants are formulated with many performance additives and base stocks to match the lubrication requirements of the equipment in which they are used. Some lubricants are incompatible with each other because of differences in additive chemistry that lead to undesirable chemical reactions. If these oils are mixed, insoluble material may form and then deposit onto sensitive machine surfaces.

This may lead to mechanical problems and shorter equipment life. Also, if incorrect lubricant are used, it may not develop an adequate film thickness and may cause equipment damage and health and safety issues. Proper labeling helps ensure that the appropriate lubricant is always used in the right component.

Color-coded labels may facilitate product identification. The fading or changing of label occurs over time. Hence color-coded labels may be combined with an alphanumeric system. Whatever labeling system you develop at your facility, it should be easy to understand and maintained by the operators and technicians. The labeling must be used consistent with all the lubricants stored at your facility. If lubricants are stored outside, use labels that resist moisture and sun damage.

HEALTH AND SAFETY PRECAUTIONS

In a typical lubricant storage facility, there are many sources of safety issues. Potential injury may occur during transportation and storage of lubricants. Forklifts carrying oil drums, when moving in confined spaces, pose safety risks to pedestrians as well as the machine operators. Containers such as oil drums weighing nearly 180 kg (397 lb) could fall or roll onto unsuspecting individuals. Slippery lubricants may leak or spill, creating slip-and-fall and fire hazards. In addition to safety concerns, lubricant spills also may present environmental hazards. Proper containment, cleaning, and disposal are critical in the event of a spill.

Preparation for Worst-Case-Scenario Spills

Maintain enough oil-absorbing products (such as pads and granules) to absorb all the oil from your facility's largest container or stack of containers. If a spill or leak occurs, once most of the fluid has been absorbed and removed, clean the floor with a solvent or degreaser.

The floor of the storage facility must be clean, dry, and free of clutter. Anything that is not in the right place, such as tools, equipment, lubricant containers, empty cartons, rags, and trash, should be properly stored or disposed of.

Drums must not be allowed to drop or fall on their sides. Forklift or other equipment such as hydraulic lifts or secured skids should be used to lower drums, and at least two people are needed to lay a drum on its side from an upright position. When rolling drums, never let them roll freely on their own moment.

Sometime inhalation of vapors or oil mists and frequent and prolonged contact with mineral oils may cause health issues. Care must be taken to keep the oil off the skin and away from the eyes and to avoid ingestion or inhalation of vapors. Operators are advised to follow the basic general health and safety precautions, such as wearing working overalls, impermeable aprons, and gloves to eliminate unnecessary contact with oil.

FIRE PRECAUTIONS

Packaged lubricating oil and grease do not represent a serious fire hazard, but lubricating oil is potentially dangerous in conjunction with more flammable materials. Oil-soaked sawdust, rags, or cleaning paper, if accumulated near a heat source, may catch fire. If soaked with fatty oils, they can ignite simply by coming into contact with a high-temperature steam pipe. Most lubricants have the potential for combustion and explosion in certain circumstances. The hazard is related to the flash point of the lubricant. Lubricants with a flash point of less than 55°C (131°F) should be stored in closed containers away from heat in a well-ventilated place. Products with a flash point of 55°C (131°F) or greater require no special fire precautions but should be stored away from heat whenever possible. Maintain current Material Safety Data Sheets (MSDSs) for all stored lubricants and other chemicals, and follow all safety guidelines included in those documents. Stores where oil drums are kept must be equipped with CO_2, dry-chemical, or foam-type fire extinguishers and with sand-filled fire buckets.

Contamination Control

The easiest way to control contamination is to avoid using practices that have a risk of exposing the lubricant to a contaminant(s). Contamination of newly commissioned storage and handling equipment includes preservatives, paint, moisture, rust particles, and fabrication debris such as dust and dirt. Before storage and handling, the equipment should be thoroughly cleaned, made scale free, and treated internally with a protective coating that is compatible with the lubricants used. The equipment should be carefully dried and cleaned before it is charged with a lubricant to be stored. Lubricants in storage are most prone to become contaminated with water from headspace condensation. Dissolved, emulsified, and free waters pose potential risks to lubricating oil.

PROPER LONG-TERM STORAGE

Lubricants degrade mainly because of prolonged exposure to heat and contamination. To avoid lubricant deterioration, it should be stored for only a limited period of time. The amount of time that lubricants can be stored depends on various factors such as base oil quality, type of additive(s), type of thickener(s), storage temperature, variation of temperature, humidity, type of container (metal drums or plastic container), and outdoor storage.

Certain additives by design will never dissolve with base oil. For example, some gear oils may be formulated with solid additive suspensions such as graphite and molybdenum disulfide. These oils should not be stored for prolonged periods because the solid additives are prone to settle in the tank.

Long-term storage at moderate temperatures may have little effect on certain lubricating oil, but at higher temperatures, oxidation occurs in all oils that are in contact with air. The quality of the base oil and additives also affects the rate of oxidation. Increasing the temperature at which the lubricant is stored by 10°C (18°F) doubles the oxidation rate, which cuts the usable life of the oil to half. The presence of water and contamination usually introduced by condensation and temperature variations increases the rate of oxidation.

Some products may deteriorate and become unsuitable for use if stored longer than three months to a year from the date of manufacture. Table 9.1 provides a general guideline for the maximum amount of time a lubricant should be stored to avoid performance degradation in normal conditions (clean and dry) and temperatures [4–38°C (40–100°F)]. If a product exceeds its maximum recommended storage time, it should be sampled and tested to confirm fitness for purpose.

TABLE 9.1 Maximum Storage Times for Various Lubricants

Product	Maximum Recommended Storage Time, Months
Lithium grease	12
Calcium complex grease	6
Gear oils	6
Fluids or lubricants with solid additives	3
Turbine oils, hydraulic fluids, R&O oils	18

Handling of Lubricating Oil

Although all care is taken during storage of lubricants, often lubricants get contaminated between storage and end use in the equipment. Many root causes of contamination causing machine shutdown are mainly due to a lack of attention in the distribution of a lubricant.

Even though the right lubricants are used consistently, many cross-contaminations occur because the same dispensing equipment is used for multiple lubricants. Ideally, dispensing equipment (such as pumps, transport containers, funnels, grease guns, filter carts, and transfer carts) should be labeled using the same system applied to the lubricants, matching each piece of equipment to the lubricant with which it is to be used. The same labeling system should be applied to machine lubrication points as well. Plastic jugs are preferred. Metal jugs can rust. Galvanized steel jugs can result in the transfer of zinc into the oil, and zinc can be poisonous with respect to some machine parts (e.g., turbine plain bearings). It is preferable to use jugs with a lid to prevent pollution by dust between storage and use in the equipment. When the jugs are not being used, cover them with a rag to prevent pollution by dust (Figure 9.7). Use a different grease gun for each type of grease. The nozzles of the grease guns must be wiped off after use. Grease nipples on the machine must be wiped off before packing them with grease.

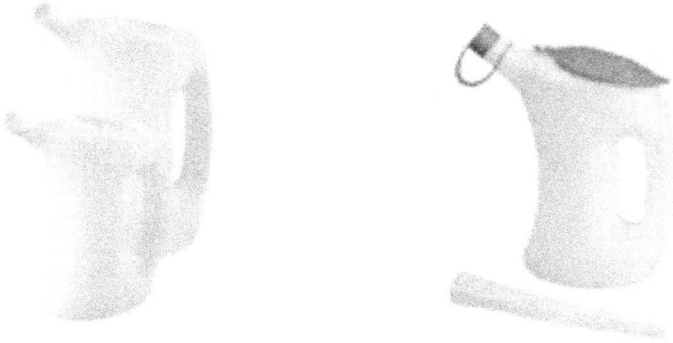

FIGURE 9.7   Dispensing equipment for lubricants.

Transferring Large Volumes of Oil

When adding large volumes, prefer the use of a vacuum pump with a filtration unit. In the same way as for jugs, it is better to use a different pump for each grade of oil. Otherwise, flush the pipes out thoroughly, and replace the oil filter before topping up again (Figure 9.8).

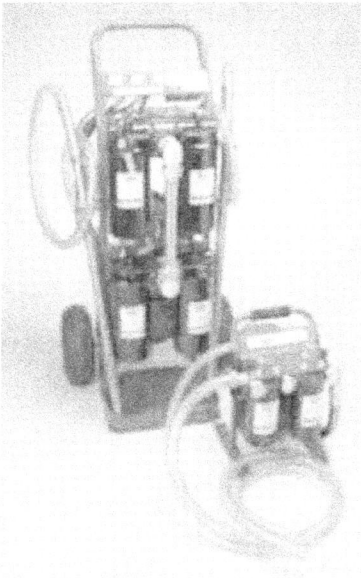

FIGURE 9.8   Lubricant prefiltration pump.

PREFILTERING OF LUBRICANTS

Whenever the concentration of solid contaminants in the lubricant exceeds the QC target limits for allowable contaminants, the lubricant should be filtered before use. Filtering lubricants prior to use can be done easily and relatively inexpensively. Filter elements should have high dirt-holding characteristics, and low backpressure limits.

USE OF SEALABLE, CLEANABLE TOP-UP CONTAINERS

After the effort of testing and prefiltering the products to ensure that they are capable of performing in the desired manner, the end user certainly wants to preserve the improved quality state of the lubricant while it is being transported around the plant in the possession of the lubrication technician. This can be challenging. Any surface that is wetted with oil becomes a magnet for atmospheric contaminants. Funnels, new oil cans, dipsticks, transfer container, and so on all eventually become grossly contaminated if they are not isolated from the atmosphere following use. For items small enough to fit in a plastic bag, such as a funnel, once used, the items should go immediately into a Zip-Loc-type resealable plastic bag. Open metal cans should be replaced with oil-handling containers designed for this purpose.

A number of sealable and cleanable oil-handling containers are available to transport the prefiltered lubricant to the machine center. Discharge nozzles range in size, opening diameter, and type from pump-type quarter-inch nozzles to gravity-flow, 1-inch, twist-to-seal-type nozzles. The containers also come in a variety of colors, as shown in Figure 9.9.

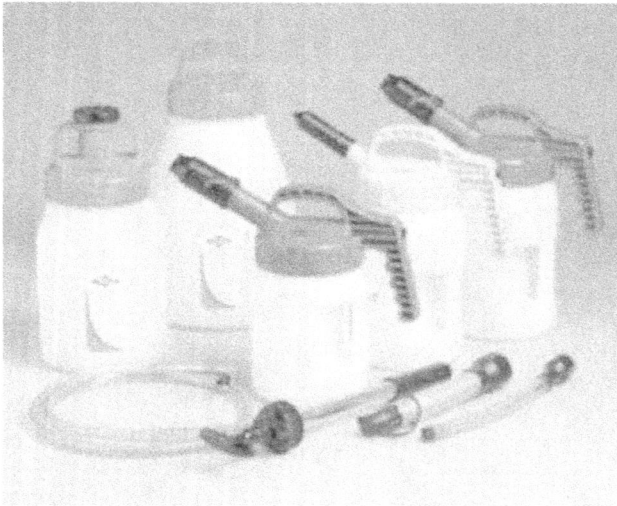

FIGURE 9.9   Sealable and cleanable oil-handling containers with color coding.

PRECAUTIONS DURING TOP-UP ACTIVITIES

Provided that the lubricant is delivered according to quality expectations and maintained in a clean state while in inventory, the remaining opportunity for corruption occurs at the time that the lubricant is placed into the machine sump. The act of filling the machine is the last chance to inadvertently harm the lubricant and the machine. A few simple precautions can eliminate the remaining threat.

When topping reservoirs with a high volume of lubricant (10 gallons or more), equip the sump drain and fill ports with fluid quick connectors that, matched with the quick connectors on the filter cart, allow the prefiltering process to take place from the lubricant package directly into the lubricant sump. The amount of precleaning of the fitting necessary before each use is simple and quick if the equipment owner keeps the connector covered with a rubber or metal cap while the machine is in normal running mode. Observe the same consideration for flow rate and filter quality as was discussed previously.

When topping reservoirs with a low volume of lubricant (10 gallons or less), follow these guidelines:

1. Use a soft bristle brush to completely displace any atmospheric dust or dirt that has accumulated around the port plug since the last top-up.
2. If there is wet residue, use a clean, lint-free cloth to physically wipe down the area around the port plug.
3. Loosen the plug and repeat step 1 if solid debris is present.
4. Remove the plug and place it in a clean and dry container or location while topping the port.
5. Reinspect and try to remove any contaminant that may fall into the sump during top-up.
6. Use a top-up reservoir that has a piston-type displacement pump supplying lubricant through a narrow-opening discharge fitting. Place the fitting into the reservoir, and fill the sump to the appropriate level.
7. Replace the plug, and wipe down any residual oil on the housing.
8. Drain any excess, if necessary. This is not clinical-type work but simple precautions to minimize the kind of ingression that can greatly support the long-term reliability objectives for the production machinery.

Are the Lubricants Ready for Dispensing?

Lubricants in storage are not as idle as they may appear. There is activity inside those pails, drums, tanks, and other vessels. Additives in stored lubricants may settle over time, so they might not be evenly distributed when the lubricant is needed. Also, as discussed earlier, air, moisture, and other contaminants may find their way into the fluid.

For these reasons, prior to dispensing, containers of stored lubricants should be agitated (or circulated with a circulating rig, in the case of bulk containers) to help ensure even, consistent distribution/concentration of additives throughout the fluid. Stored lubricants also should be filtered before dispensing.

APPROPRIATE STOCK ROTATION

To avoid shelf-life issues for petroleum products, the lubricant inventory volume should be maintained low enough that the whole inventory is turned over at least once a year. To facilitate this, the storage area should be arranged in such a way that its contents will accommodate first-in, first-out rotation. In addition, a first-in, first-out (FIFO) stock management system is preferred so that the products stored first are used first (Figure 9.10). This will help ensure use of older lubricants first because lubricants stored for longer periods may deteriorate.

In   Out

FIGURE 9.10   FIFO storage management.

CONCLUSION

Some of the most common causes of lubricant contamination and degradation include unclean dispensing equipment, mixing of lubricants, damaged or improperly sealed containers, ingression of moisture from humidity, condensation or precipitation, a dusty or dirty environment, exposure to extreme heat or to fluctuating temperatures, and storage beyond a lubricant's expected shelf life. Precautions should be taken to safeguard each of these storage and handling issues and develop best practices to get optimal performance, protection, and service life of your lubricants.

ICML Questions

1. Which of the following is (are) a storage and handling issue for lubricants?
   a. Dust and water may enter the lubricant if adequate precautions are not taken.
   b. Damaged packaging can result in leaks.
   c. If the lubricants are not labeled properly, an unsuitable lubricant may be added to a machine.
   d. All of the above

2. Lubricant drums are best stored on their sides (horizontally) to prevent moisture from accumulating on the top
   a. with bungs at the three and nine o'clock positions.
   b. with bungs at the six and twelve o'clock positions.
   c. with bungs at the five and eleven o'clock positions.
   d. with bungs in any position.

3. Most good-quality synthetic and conventional mineral oils are not affected by storage temperatures
   a. below 49°C.
   b. below 100°C.
   c. below 75°C.
   d. below 80°C.

4. Common causes of lubricant contamination and degradation include
   a. unclean dispensing equipment.
   b. mixing of lubricants.
   c. intake of moisture from humidity.
   d. All of the above

5. The maximum storage limit for lithium grease is
   a. 4 months.
   b. 6 months.
   c. 8 months.
   d. 12 months.

CHAPTER 10

Oil Sampling

One of the most important milestones of an oil analysis program is sampling of the oil. The way a sample is collected, the frequency, the accessories used, and the procedures followed all dictate how informative the oil samples will be and, subsequently, dictates how beneficial the results will be. This is why it is of great importance to take the sample correctly. If the oil sample is not representative of the oil in the machine, then the results will not accurately reflect the condition of that component. The result will be an incorrect diagnosis based on incorrect analysis because of a poorly taken sample. Establishing effective, user-friendly oil-sampling procedures helps to build an oil analysis program that creates value through better maintenance decisions.

Objectives of Strategic Oil Sampling

- Maximize data density
- Minimize data disturbance
- Maintain proper frequency

Strategic Sampling Considerations

- Sampling location
- Sampling hardware
- Sample bottles
- Sample procedure

DATA DENSITY AND DATA DISTURBANCE

The oil sample should represent the body of oil about which data are required in order to increase the effectiveness of oil and machine decisions. This will happen by maximizing data density and minimizing data disturbance. Taking a sample that maximizes data density and minimizes data disturbance is of vital importance when selecting a sampling location.

The first goal is to maximize data density. The sample must be taken in such a way as to ensure that there is as much information per milliliter of oil as possible. This information relates to such criteria as cleanliness and dryness of the oil, depletion of additives, and the presence of wear particles generated by the machine. However, maximizing data density depends on the nature of the data you desire. For example, if you want to assess the effectiveness of a system's filter, you must collect a representative sample before and after filtration. The difference between the two samples is reflected in the differential particle count across the filter. Depending on the results, a decision to retain or change the filtering practices can be made.

In this instance, maximizing the density of information requires the analyst to obtain two representative samples from a specific location to calculate the information required. Different objectives require different sampling procedures. The objective should drive the sampling procedure.

The preceding example applies to transient properties such as measuring particles (wear and contaminants). Transient properties depend on the location from which the sample is collected. By contrast, homogeneous properties such as viscosity, total acid number (TAN), and total base number (TBN) tend to remain constant throughout the oil. Transient properties pertain to equipment health and contamination, whereas homogeneous properties pertain to oil and additive health. It is more difficult to maximize the data density of transient properties, and for this reason, effective sampling of transient properties is essential if reliable and trendable data are expected.

The second goal is to minimize data disturbance. The sample should be extracted in such a way that the concentration of information is uniform, consistent, and representative. It is important to make sure that the sample does not become contaminated during the sampling process. This can distort and disturb the data, making it difficult to distinguish what was originally in the oil from what came into the oil during the sampling process. Failing to sample from a running machine, where the oil is not hot or well mixed, is a common source of data disturbance. Ideally, the machine should be operating at normal load and speed in its typical environment when a sample is taken; otherwise, particles and moisture can settle before the sample is taken, causing data disturbance.

Using dirty sampling equipment and exposing open bottles and caps to the environment can disturb the quality of the data. A common example of data disturbance is the cleaning of sampling equipment and bottles with diesel fuel or solvents. Even residual amounts of diesel fuel from this cleaning can be detected by most oil analysis laboratories and mistakenly diagnosed as evidence of a fueling problem on an engine.

By communicating a potential source of interference with a given sampling method or location, the diagnostician can be on the lookout for these pitfalls, reducing the likelihood that the oil analysis results will be compromised. This is especially important when the sampling point or required process is less than ideal because of the location of the machine or operation-related restrictions. In the case of a sample bottle washed out with diesel fuel, the simplest communication by the sample taker could avert an overreaction to the fuel detected in the sample.

The oil sample should be extracted in such a way that the concentration of information is uniform, consistent, and representative. It is important to make sure that the sample is not contaminated during the sampling process. At the end of the day, you want to ensure that what enters the bottle is both rich in information and remains undisturbed by the sampling process itself.

The hardware used to extract the sample should not disturb sample quality. It should be easy to use, clean, rugged, and cost-effective. In addition, it is important to use the correct bottle type and bottle cleanliness to ensure that a representative sample is achieved.

SAMPLING FREQUENCY

Sampling at the correct frequency is vitally important, and oil analyses are usually carried out far less frequently than is ideal. Machinery manufacturers often suggest a sampling interval, but that should only be a rough guideline. The equipment owner is the best judge of sampling intervals. Pertinent questions to consider when arriving at a sampling interval include:

- Criticality of equipment (or lack of redundancy)
- Environment (is it wet, dry?)
- Operating conditions (load, speed)
- Failure history
- The cost of a failure (in repair costs, lost production, life and safety)
- Changes in operating conditions (that put more stress on the machine)
- Safety risk (potential loss of life if catastrophic failure occurs)

A rough rule of thumb is that critical equipment should be sampled at least once a month. Noncritical equipment for which oil analysis is warranted should be sampled no less frequently than once every three months. In general, a quarterly or monthly sample interval is appropriate for most important industrial machinery, whereas reciprocating engines need to be sampled at a more frequent interval based on time of oil. Having on-site oil analysis equipment allows the user to self-manage and extend intervals safely as trends develop. Further guidance may be sought by consulting with the equipment manufacturer and oil supplier.

STRATEGIC SAMPLING CONSIDERATIONS

The following questions should be answered when designing a sampling program:

- Where is the best location to draw an oil sample to ensure that the correct information is collected?
- What are the best tools for drawing a sample from a specific location?
- Who will be responsible for pulling the sample, and how consistent will the sample be each time it is drawn from the specific location?
- What are the procedures followed to collect the sample?
- How clean is the sampling hardware?

Locations for Sampling

Sampling Port Location
Troubleshooting problems in oil analysis is greatly assisted by the installation of several sampling ports in various locations to isolate individual components. Using multiple sample ports provides an analytical edge for both discovering potential component failures and analyzing the root causes. Sample ports are classified into two categories, primary and secondary.

Primary Sampling Ports.   The primary sampling port is the location where routine oil samples are taken. The oil from this sample location is usually used for monitoring oil contamination, wear debris, and the chemical and physical properties of the oil. Primary sampling locations vary from system to system but are typically placed on a single return line prior to entering the sump or reservoir.

Secondary Sampling Ports.   Secondary sampling ports can be placed anywhere on the system to isolate upstream components. This is where contamination and wear debris contributed by individual components will be found.

Consider a lube oil pump that feeds three sets of bearings (Figure 10.1). The return for the three bearings combines into a single return line before entering the sump. The primary sample port is on the single return line after all three bearing lines join and before the oil enters the sump. The secondary sampling locations are immediately downstream of the pump (upstream of all three bearings) as well as downstream of all three bearings.

For example, when performing an on-site particle count from a sample from a ferrous metal count on the primary sample port, a particle count corresponding to an ISO code of 16/13 is reported (32 percent of which are ferrous particles). This level has exceeded the target of 14/11 and reflects a problem in the system. It also shows that the contamination level of the oil returning to the sump is quite high even though the concentration may be somewhat diluted.

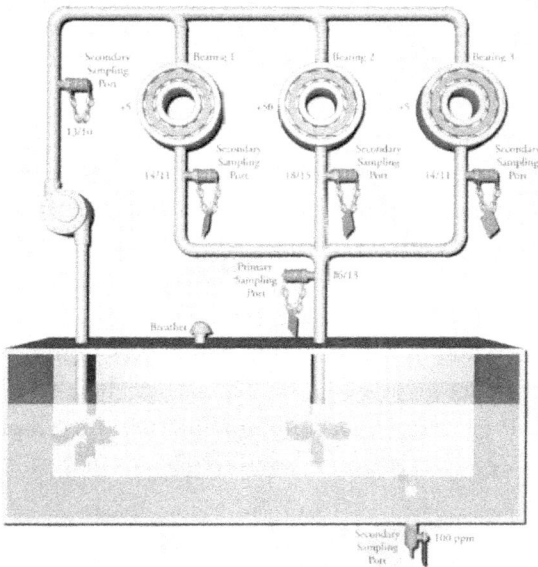

FIGURE 10.1   Sampling port locations on a lube oil pump that feeds three sets of bearings.

The sample drawn from the primary sample port informs us that something has changed in the system, but it does not show what has changed. This is why secondary sample ports are used. They pinpoint the precise location of the problem. The particle counts on the secondary ports indicate a cleanliness level of ISO 13/10 downstream of the pump. Therefore, the oil being delivered to the bearings is fairly clean, suggesting that the pump is not the cause of the wear. The secondary sample collected and tested after bearing 1 shows a cleanliness level of ISO 14/11. Bearing 2 reported an ISO codes of 18/15 with 25 percent ferrous, and bearing 3 reported an ISO code of 14/11. The tests indicate that bearing 2 is showing wear because the oil delivered to the bearings was fairly clean and below the target level of ISO 14/11. Exception testing of the oil sampled from bearing 2, such as analytical ferrography, is used to determine what is causing the wear.

Sampling multiple points on equipment takes most of the guesswork out of pinpointing troubled components. Use this method of oil sampling to schedule downtime and avoid maintenance to components that do not require replacement. Secondary sample ports also can be used to monitor the general performance of filters. The primary port will show what is going into the filter, whereas the secondary ports show what is coming out. This procedure enables a filter change based on condition, long before the differentiated pressure indicator shows that a filter is in bypass.

Sampling of Circulating Systems

When a sample is taken from a line in a circulating system, it is referred to as a *live-zone sample*. There are things that can be done during the sampling process that improve the quality and effectiveness of live-zone oil sampling. There are several rules for properly locating sampling ports on circulating systems. These rules cannot always be precisely followed because of various constraints in the machine's design, application, and plant environment. However, the rules outlined next should be followed as closely as possible.

Turbulence.   The best sampling locations are highly turbulent areas where the oil is not flowing in a straight line but is turning and rolling in the pipe. Sampling valves located at right angles to the flow path in long, straight sections of pipe can result in particle flyby, which basically leads to a substantial reduction of the particle concentration entering the sample bottle. This can be avoided by locating sampling valves at elbows and sharp bends in the flow line (Figure 10.2).

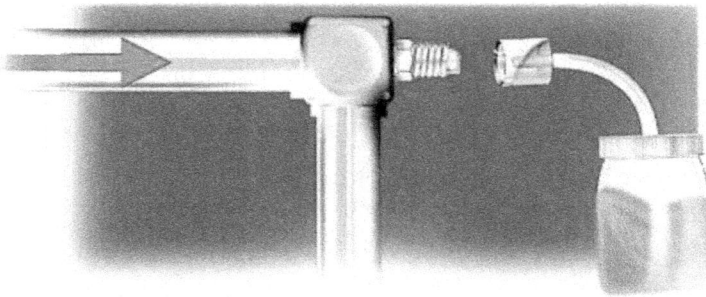

FIGURE 10.2   Highly turbulent area.

Ingression Points.   Where possible, sampling ports should be located downstream of the components that wear and away from areas where particles and moisture ingress. Return lines and drain lines heading back to the tank offer the most representative levels of wear debris and contaminants. Once the fluid reaches the tank, the information becomes diluted.

Filtration.   Filters and separators are contaminant removers; therefore, they can remove valuable data from the oil sample. Sampling valves should be located upstream of filters, separators, dehydrators, and settling tanks unless the performance of the filters is being specifically evaluated.

Drain Lines.   In drain lines where fluids are mixed with air, sampling valves should be located where oil will travel and collect. On horizontal piping, this will be on the underside of the pipe. Sometimes oil traps, such as a gooseneck, must be installed to concentrate the oil in the area of the sampling port. Circulating systems where there are specific return lines or drain lines back to a reservoir are the best choice for sampling valves (Figure 10.3).

TABLE 10.1   Dos and Don'ts of Live-Zone Sampling

Dos	Don'ts
Do take samples from turbulent zones such as elbows.	Don't take samples from dead legs.
Do take samples from live zones.	Don't take samples from laminar-flow regions.
Do take samples when the machine is running.	Don't take samples after the filter.
Do take samples when the machine is under normal operating conditions.	Don't take samples when the machine is cold and not operating.
Do take samples downstream of system components (bearings, gears).	

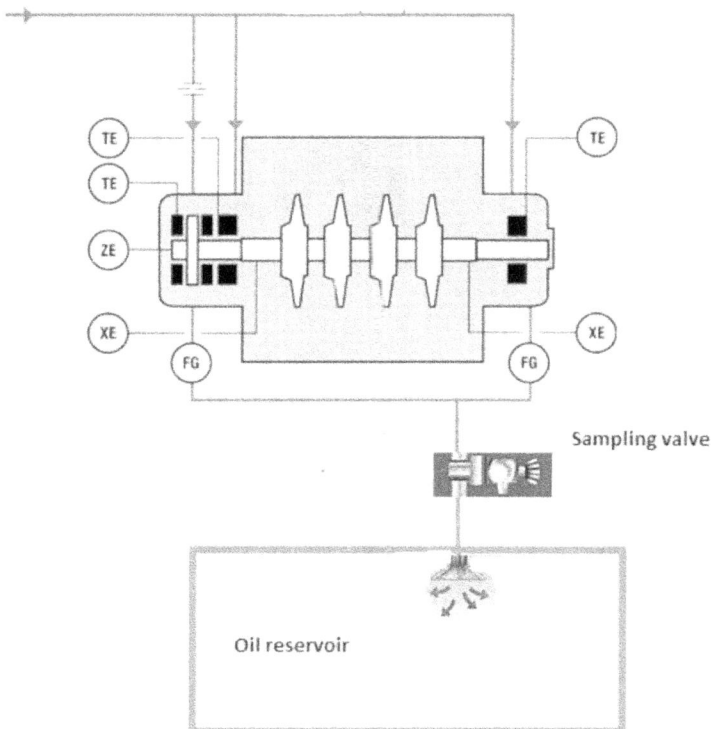

FIGURE 10.3   Return or drain line.

Circulating systems with specific return or drain lines allow the sample to be taken before the oil returns to the tank and always before it goes through a filter. If the oil is allowed to return to the tank, then the information in the sample becomes diluted, potentially by thousands of gallons of fluid in large lubricating and hydraulic systems. In addition, debris in the reservoir tends to accumulate over weeks and months and may not accurately represent the current condition of the machine.

Sampling Wet Sumps

Frequently, there are applications where a drain line or a return line cannot be accessed or no such line exists; these are typically called *wet-sump systems*. Examples of wet-sump systems are circulating gearboxes and circulating compressors. In these applications, because there is no return line, fluid sampling is from the pressurized supply line leading to the gearing and the bearings (Figure 10.4). The sample should be collected before the filter, if one exists.

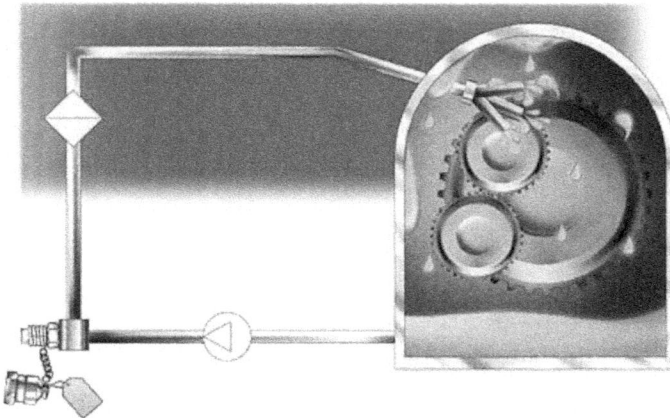

FIGURE 10.4   Pressure or feed line.

The best place to sample crankcase oil is also just before the filter. The sampling valve should be installed between the pump and filter. This sample location is highly preferred over sampling from a drain port or using a vacuum pump and tube inserted down the dipstick port. Many newer model engines come with an appropriately located sample valve right on the filter manifold.

Sampling Noncirculating Systems

There are numerous examples where no forced circulation is provided, and a sample must be taken from a system's sump or casing. This occurs often with *in-service equipment* on the run.

Oil bath–lubricated bearings and splash-lubricated gearboxes are common examples of these systems. All of these situations increase the challenge of obtaining a representative sample.

The most basic method for sampling such sumps is to remove the drain plug from the bottom of the sump and allowing fluid to flow into the sample bottle. For many reasons, this is not an ideal sampling method or location. Most important is the fact that bottom sediment, debris, and particles (including water) enter the bottle in concentrations that are not representative of what is experienced near or around where the oil lubricates the machine. The drain-plug sampling method should be avoided if at all possible.

Sumps and reservoirs were designed to hold a large volume of oil, to dissipate heat, and to allow air to rise and contaminants to settle. Therefore, the most concentrated contamination is on the bottom of the sump or reservoir, and the cleanest oil toward the top. The ideal location for drawing an oil sample from a sump or reservoir is at 50 percent of the oil level.

Drain-port sampling can be greatly improved by extending a short length of tubing inward and up into the active moving zone of the sump. Ideally, the tip of the tube, where the oil sample is taken, should be halfway up the oil level, 2 inches from the walls, and at least 2 inches from the rotating elements within the sump (Figure 10.5).

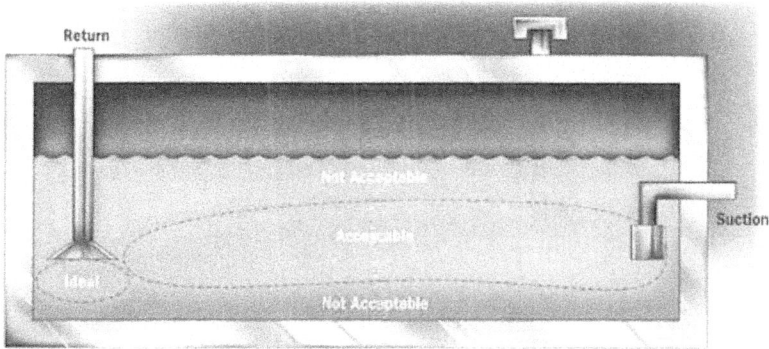

FIGURE 10.5 Appropriate level for oil sampling from a sump or reservoir.

If the drain port is the only way to obtain a sample from a gearbox, there are commercially available sample tubes that can be installed on the bottom or side of the sump (Figure 10.6). These inward pilot tubes can be manipulated to ensure that the sample is drawn from the most appropriate location of the sump or reservoir and that the sample is taken from the exact same location inside your system each time. This method is a more consistent and representative way of sampling oil than drop-tube sampling (Figure 10.7).

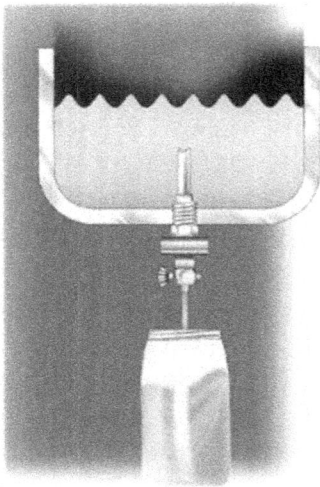

FIGURE 10.6   Sample tube installed on the bottom of a sump.

FIGURE 10.7   Drop-tube sampling.

OIL SAMPLING TOOLS

A comprehensive and well-managed oil sampling process will provide the correct samples of lubricants for analysis. Taking bad samples can have disastrous consequences. Using the wrong tools, taking oil samples from the wrong parts of the lubrication system, taking the samples improperly, or handling the samples incorrectly will yield a sample that does not truthfully reflect the condition of the system. This will cause the oil analysis to return incomplete or incorrect results.

Sampling Valves/Ports

Sample ports are used for online sampling and monitoring of system fluids without system shutdown. The ports incorporate a check valve for reduction of fluid contamination that is normally closed until opened by the sampling adapter for drawing a sample. They are equipped with a protective dust cap to eliminate ingression of contamination into the system.

Pitot Sampling Tubes/Probes

Pitot sampling tubes are designed to provide a safe, simple, and effective method of sampling fluids from sumps and nonflooded horizontal drain lines. They ensure that oil samples are drawn from the most appropriate location of the sump reservoir and that the sample is taken from the exact location inside the system each time. This is important for maintaining consistency in routine sampling (Figure 10.8).

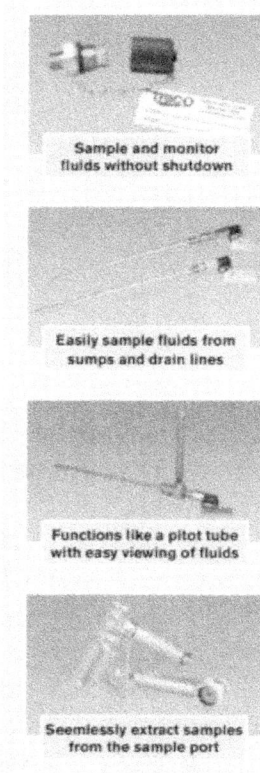

FIGURE 10.8   Sampling instruments.

Liquid Level Gauge Sampling Ports

Liquid level gauge sample ports offer the same sample benefits as pitot sampling tubes, but they also provide easy viewing of the fluid level and condition. They are ideal for bearing housings and other nonpressurized applications.

Vacuum Pump

A vacuum pump is a necessary tool for extracting an oil sample from a sample port. Used in combination with a sample-port adapter, flexible tubing, and a sample bottle, it allows users to connect to any sample port for contamination-free oil sampling.

SAMPLING BOTTLES AND HARDWARE

An important factor in obtaining a representative sample is to make sure that the sampling hardware is completely flushed prior to obtaining the sample. This is usually accomplished by having a spare bottle to catch the purged fluid. It is important to flush 5–10 times the dead-space volume before obtaining the sample. All hardware with which the oil comes into contact is considered dead space and must be flushed, including:

- System dead legs
- Sampling ports, valves, and adapters
- Probes on sampling devices
- Adapters for using vacuum pumps
- Plastic tubing used for vacuum pumps (this tubing should not be reused to avoid cross-contamination between oils)

An assortment of sampling bottles is available for used in oil analysis. An appropriate bottle needs to be selected for the application and test that is planned. Several features, including size, material, and cleanliness, must be considered when selecting a sample bottle.

A number of different-sized sampling bottles are available. They vary in size from 50 ml to the more common 100–120 ml. The larger bottle is preferred when tests such as particle count and viscosity analysis are required. Where a considerable number of different tests are required, a 200-ml bottle (or two 100-ml bottles) may be required. It is important to coordinate with the laboratory when selecting a bottle size that will provide sufficient volume to conduct all the required tests and leave some extra for storage in case a rerun is necessary.

Another consideration in selecting bottle size is that the entire volume of the bottle should not be filled with fluid during the sampling process. Only a portion of the sampling bottle should be filled. The unfilled portion, called the *ullage*, is needed to allow proper fluid agitation by the laboratory to restore even distribution of suspended particles and water in the sample. The general guidelines for filling bottles include:

- **Low viscosity (ISO VG 32 or lower).** Fill to about three-fourths of the total volume.
- **Medium viscosity (ISO VG 32 to ISO VG 100).** Fill to about two-thirds of the total volume.
- **High viscosity (over ISO VG 100).** Fill to about one-half of the total volume.

Bottles are available in several materials. Plastic polyethylene is one of the most common bottle materials. It is an opaque material similar to a plastic milk jug. This type of sampling bottle presents a drawback because the oil cannot be visually examined after the sample is obtained. Important oil properties, such as sediment, darkness, brightness, clarity, and color, can be immediately learned from a visual inspection of a clear bottle.

Another material is PET plastic. It is a completely clear, glass-like material and is available in standard-sized bottles. This plastic is compatible with most types of lubricating oils and hydraulic fluids, including synthetics.

Of course, glass bottles are also available. These bottles tend to be more expensive and are heavier, and there is the risk of breakage during the sampling process. One advantage of glass bottles is that they can be cleaned and used over and over. The cleanliness of glass bottles often exceeds that of plastic bottles. One of the most important considerations in selecting a sampling bottle is to make sure that it is sufficiently clean.

CONCLUSION

The efficacy of a sampling program depends on many factors. There should be consistency within the program. Procedures and guidelines for the program should be set up and monitored. Monitoring should include the frequency of sampling on a given piece of equipment, the amount of oil to flush prior to pulling the sample, the method by which the oil sample is attained, the tools used to get the sample, how the bottle is to be labeled and what information it will contain, which tests are done for a specific machine on a regular basis and which tests are done on exception, and anything else that will add to the integrity of the sample being analyzed. Approaching oil sampling with knowledge and creativity will allow a program to reap the available benefits.

ICML Questions

1. What are the objectives of strategic sampling?
   a. Maximize data density
   b. Minimize data disturbance
   c. Maintain proper frequency
   d. All of the above

2. Sampling data disturbance refers to
   a. extracting samples so that the concentration of information is uniform, consistent, and representative.
   b. not contaminating samples during the sampling process.
   c. failing to sample from a running machine.
   d. All of the above

3. In sampling of medium viscosity (ISO VG 32 to ISO VG 100) oil,
   a. the bottle should be filled about two-thirds of the total volume.
   b. the bottle should be filled about three-fourths of the total volume.
   c. the bottle should be filled about one-half of the total volume.
   d. any volume is acceptable.

CHAPTER 11

Understanding Oil Analysis: How It Can Improve Reliability

The primary function of a lubricant is to *lubricate* and reduce friction between surfaces. When a proper lubricant is used and a proper load is applied, the asperities are not in contact, and no wear will occur. However, when the lubrication is inadequate or the load is increased, the oil film will not be thick enough to fully separate the asperities. The largest asperities will come in contact, resulting in increased wear. Sometimes particulate contamination leads to increased wear through abrasion and reduced oil flow. Oil analysis is performed to determine both fluid and machine integrity by measuring the elements found in the four main categories: physical fluid properties, fluid contaminants, additive metals, and wear debris. This process involves analyzing a lubricant's chemical and physical properties in order to detect lubricant or equipment issues before major problems develop. Hence oil analysis serves as a valuable preventative maintenance tool, allowing an accurate scientific look at the lubricant's service condition as well as the condition of the operating equipment.

This chapter provides an introduction to oil analysis of used lubricating fluids. It presents insight into the basic principles and test methodologies used in this field. The four primary focus areas of oil analysis tests reveal information about lubricant condition, contamination, additive elements, and wear debris.

1. **Lubricant condition.** The condition of the lubricant is monitored with tests that quantify the physical properties of the oil to ensure that it is serviceable. Metals and debris associated with component wear are measured to monitor equipment health. Lastly, some tests target specific contaminants commonly found in oils. Assessment of the lubricant condition reveals whether the system fluid is healthy and fit for further service or is ready for a change.

2. **Contaminants.** Increased contaminants from the surrounding environment in the form of dirt, water, and process contamination are the leading cause of premature machine degradation and failure. Increased contamination alerts you

to take action to save the oil and avoid unnecessary machine wear. Similarly, the presence of grease contaminating an oil system may be indicated by the increased presence of aluminum or barium if the grease contains metallic soaps. Although contamination is commonly associated with substances entering a component's oil system from an outside source, wear metals themselves are also a form of contamination.

3. **Additive elements.** These are substances that have been added to the oil to impart particular characteristics needed for specific applications. A measure of additive elements can provide the necessary information to determine whether the oil is still chemically able to perform its job. For example, the additive calcium (Ca) is used as a detergent; if oil analysis shows calcium levels to be too low, it would be correct to surmise that the oil will not be able to prevent sludge and deposits. The additive metals category focuses on chemical properties of the oil and compares them with baseline metals previously established for the oil.

4. **Machine wear.** An unhealthy machine generates wear particles at an exponential rate. The detection and analysis of these particles assist in making critical maintenance decisions. Machine failure as a result of worn-out components can be avoided. Remember, healthy and clean oil leads to the minimization of machine wear.

Wear debris/wear metals are elements that are introduced into the oil as a result of the wearing down of machine components. Most commonly, these components are tiny pieces of metal, which is why they are often referred to as *wear metals*. The term *wear debris* is used interchangeably with wear metals and includes materials such as molybdenum that also may be incorporated into machine systems. Combinations of wear debris can identify components within the machine that are wearing. For a proper analysis of wear debris/metals, the component materials incorporated into the particular system must be known.

Results from wear-debris testing can indicate whether components in the system are operating in a normal state, are nearing failure, or have already failed. This is accomplished not only by assessing the type of debris/metals present but also by looking at their relative concentration in the sample, which can be indicative of specific component failures. It is imperative to select the proper blend of tests to monitor the machine's lubricant condition, wear debris, and contaminants in order to meet the goals of successful oil analysis.

OIL ANALYSIS TESTS AND THEIR SIGNIFICANCE

Without a working knowledge of oil analysis tests and their significance, the user may be uncertain about the value of the service and how each test interrelates with the others to provide a useful, accurate picture of internal component and lubricant conditions. The following information is provided as a general orientation to what analysts consider to be the most important oil analysis tests for industrial and nonindustrial applications.

Physical Analysis

Viscosity

Viscosity is a lubricant's internal resistance to flow at a given temperature in relation to time and is considered to be the single most important physical property of a lubricant. If a lubricant does not have the proper viscosity, it cannot perform its functions. If the viscosity is not correct for the load, the oil film cannot be established at the friction point. Heat and contamination are not carried away at the proper rates, and the oil cannot adequately protect the machine. Viscosity is most commonly determined with a kinematic method, and the results are reported in centistokes (cSt) The most common technique for measuring an oil's viscosity is outlined in ASTM D 445 and is accomplished using a viscometer.

Industrial oils are identified by their ISO viscosity grade (ISO VG). The ISO VG refers to the oil's kinematic viscosity at 40°C (104°F). When an oil's viscosity increases, it is usually due to oxidation, degradation, or contamination. This is the result of extended oil drain intervals, high operating temperatures, or the presence of water or another oxidation catalyst. Increased viscosity can also be the result of excessive contamination with solids such as soot or dirt, as well as topping off with a higher-grade lube. Water contamination also can cause high viscosity. A decrease in the oil's viscosity is most commonly due to contamination with fuel or a solvent. An oil's viscosity also can be affected if the wrong oil is used for top-off or replenishment.

Neutralization Number

Both the acid content and the alkaline content of a lubricant may be measured and expressed as a neutralization number obtained from a wet chemical titration.

Acid Number

Acid number (AN) is an indicator of the total amount of acidic material present in the lubricant. It is useful in monitoring acid buildup in oils resulting from depletion of antioxidants. Oil oxidation causes acidic by-products to form. High acid levels can indicate excessive oil oxidation or depletion of the oil additives and can lead to corrosion of the internal components. Based on monitoring of the acid level, the oil can be changed before any damage occurs. When the test oil is flagged for a sudden increase in acid levels, this indicates accelerated oil oxidation, and you should change the oil as soon as possible. If any of the remaining highly acidic oil is left, it will quickly deplete the antioxidants in the new oil.

AN is measured by titration based on ASTM D 664 or D 974. The results are expressed as a numeric value corresponding to the amount of the alkaline chemical potassium hydroxide required to neutralize the acid per one gram of sample.

The AN of a new oil will vary depending on the base oil additive package. An R&O oil will usually have a very low AN, around 0.03. An AW or EP oil will have a slightly higher value, typically around 0.5. Engine oils commonly have a higher AN in the neighborhood of 1.5.

Base Number

Base number (BN) measures the total alkaline content present in the lubricant. Many oils are fortified with alkaline additives to neutralize acids that are formed as a result of combustion. In diesel engine applications, acid is formed in the combustion chamber when moisture combines with sulfur under pressure. Measuring the BN will help ensure that a sufficient amount of additives have been added to the oil to help resist oxidation from acid.

Abnormal decreases in BN indicate a reduced acid-neutralizing capacity and/or a depleted additive package. The test first determines the amount of acid required to neutralize the alkaline content of the sample. The final result is then expressed as an equivalent amount of the alkaline potassium hydroxide per gram of sample. The BN of oil is highest when the oil is new and decreases with use. Once again, condemning limits are based on the application. As a rule, the BN should not drop below half its original value. BN values for new engine oils vary greatly depending on the application.

The testing is very similar to AN testing except that the properties are reversed. The sample is titrated with an acidic solution to measure the oil's alkaline reserve. ASTM D 2896 and ASTM D 4739 outline the most commonly used methods to measure the BN.

Water Contamination

Water contamination is detrimental to any lubricant. When free water (nonemulsified) is present in oil, it poses a serious threat to the equipment. Water is a very poor lubricant and promotes rust and corrosion of the components. Dissolved water in oil (emulsified) will promote oil oxidation and reduce the load-handling ability of the oil. Water in any form will cause accelerated wear, increased friction, and high operating temperatures. If left unchecked, water will lead to premature machine failure. Excessive levels of water promote lubricant breakdown and corrosion of machine components. In most systems, water should not exceed 500 ppm. A simple crackle test is used to determine whether water is present in oil. If a crackle test is positive, further testing is needed to quantify the amount of water by using Karl Fischer titration based on ASTM D 6304. A measured amount of oil is introduced into a titration chamber. This solution is titrated with Karl Fischer reagent to a specific end point. The amount of reagent used and the sample volume are calculated and converted to parts per million (percent by mass).

Low levels of water (<0.5 percent) are typically the result of condensation. Higher levels can indicate a source of water ingress. Water can enter a system through seals, breathers, hatches, and fill caps. Internal leaks from heat exchangers and water jackets are other potential sources. These must be addressed suitably.

Demulsibility in lubricating oils can be defined as the ability of an oil to release water. This property is of the utmost importance when the equipment is operating in humid climates or in industrial plants with water-intensive processes such as the pulp and paper industry, the steel rolling industry, food industries, and so on. Lubricating oils have some degree of hygroscopicity, that is, the ability to absorb water; thus it would seem natural that water should be retained in the mass of some types of lubricating oils. The water that contaminates the lubricating oils can come from the atmosphere that penetrates the equipment through the vents during volume contraction and expansion with temperature or coming from the water jets of the hoses of the machine operators and must be removed. When the demulsibility of the lubricating oil decreases, it generally becomes cloudy or foamy, and there may be rapid wear of moving parts. A sample of lubricating oil may be sent to the laboratory for analysis to determine its demulsibility, using such tests as indicated in ASTM D 1401-02 or DIN ISO 6614: 2003.

Significance

Demulsibility testing is crucial for critical equipment that may be at risk for water contamination. The inability of a lubricant to separate from water will result in fluid degradation, corrosion of components, and potential failure from improper lubrication.

Applications

Demulsibility testing is commonly used for lubricants from systems that have a risk of frequent or large-scale water contaminations to ensure adequate ability to separate the water from the lubricant.

Particle Count

Particulate contamination has negative effects on all types of equipment. Particle-count testing is a way to monitor the level of solid contamination in an oil. This test uses instruments with special detectors that count and size particles present in the fluid. Results are reported as numbers of particles in a specific size range per given volume of sample.

Analysts uses two different methods to perform the particle count: the pore blockage/flow decay method and the automated particle count (APC), a laser-based method. Each has particular strengths and appropriate applications. For these methods, three size ranges represent the current ISO 4406 standard: >4 µm, >6 µm, and >14 µm.

Results from the particle count are then used to indicate fluid cleanliness via ISO classification codes. The ISO class code is expressed as three separate numbers (e.g., 20/15/12). The first number represents the relative contamination level for the first size range, and the second and third ranges are similarly calculated. Abnormal particle contamination levels are associated with increased wear, operational problems with close-tolerance components, fluid contamination or degradation, and loss of filter efficiency. Generally, the lower size ranges are considered indicative of contamination with silt, whereas the larger size ranges point to wear problems.

PARTICLE COUNT REPORTING AND ISO REGULATIONS

The International Standards Organization (ISO) developed a universal system for representing particle concentrations within a sample. The classification is used to determine oil cleanliness by identifying the amount and size of particles found in a system. This information can be important when making recommendations for equipment that exhibits sensitivity to small particles such as hydraulic and certain gear and turbine applications. Particle count is the measurement of all particles that have accumulated within a system, including metallic and nonmetallic particles, fibers, dirt, water, bacteria, and any other kind of debris. It is most useful in determining fluid system cleanliness in filtered systems including hydraulics, turbines, compressors, auto/power-shift transmissions, recirculation systems, and filtered gear systems with a fluid viscosity of less than ISO 320. Current ISO particle counts are determined at three size scales: >4 µm, >6 µm, and >14 µm per 1 ml of fluid.

TABLE 11.1 Suggested Cleanliness Levels

Low pressure—manual control 0–5,000 psi	20/18/15 or better
Low to medium pressure—electro-hydraulic controls	19/17/14 or better
High pressure—servo controlled 1,500 psi and above	16/14/11 or better
Gear pump	19/17/14
Piston pump/motor	18/16/13
Vane pump	19/17/14
Directional valve	19/17/14
Proportional control valve	18/16/13
Servo valve	16/14/11

ISO particle counts are reported using the scale in Table 11.1 and are reported as follows: 16/15/9. This result indicates that in 1 ml of fluid, there were 16 particles larger than 4 µm in size/15 particles larger than 6 µm in size/9 particles larger than 14 µm in size.

Ferrous Wear Concentration

In some cases, a particle count is not an effective test because the sample is inherently dirty, and filtering the oil may not be plausible. A particle count indicates that the sample is extremely dirty, but it does not give any indication of ferrous wear. In gearboxes, ferrous wear may be more important than overall particle count. In such an application, determination of ferrous wear concentration is a good substitution for a particle count test.

A wear particle analyzer quantifies the amount of ferrous material present in a sample of fluid. A measured amount of sample is inserted into the analyzer, and the amount of ferrous material is determined by a change in magnetic flux. This change is then converted into ferrous concentration in parts per million. One advantage of a ferrous debris monitor is that it measures ferrous wear debris in all types of oil, from gearbox lubricants through hydraulics; another key benefit is that it will also measure ferrous wear debris found in grease.

A test similar to the ferrous debris monitor is direct read (DR) ferrography. DR ferrography collects positively charged particles on two light sources and measures the amount of blocked light to determine the level of ferrous contaminants present in an oil. Although these two tests provide the same information, they are not interchangeable.

Ferrography

Ferrography is an analytical technique in which wear metals and contaminant particles are magnetically separated from a lubricant and arranged according to size and composition for further examination. It is widely used in oil analysis to determine component condition via direct examination of wear metal particles.

There are three stages in a complete ferrographic analysis: (1) direct reading (DR) ferrography, (2) analytical ferrography, and (3) ferrogram interpretation and report. DR ferrography precipitates the wear particles from a sample and electronically determines the quantity of large (>5 μm) and small (1–2 μm) particles present in the sample. Wear calculations from these results indicate the rate, intensity, and severity of wear occurring in the sampled machine. In cases where the DR ferrography wear trends indicate an abnormal or critical wear condition, analytical ferrography can reveal the specific type of wear and probable source of the wear condition.

DR Ferrography

DR ferrography magnetically separates wear particles and optically measures the quantity of large and small particles present in the oil sample. Results from DR ferrography indicate the rate, intensity, and severity of wear. With these measurements, machine wear baselines can be established, and trends in wear conditions can be monitored. If there is a significant increase in the wear trend levels, a detailed analytical ferrography should be performed. When DR ferrography indicates abnormal wear, analytical ferrography can further pinpoint its source and the specific type of wear. This analysis will extract, classify, and visually analyze wear particles and solid contaminants. Particles are examined under a powerful optical microscope to determine the size, concentration, color, shape, and composition. Results received from analytical ferrography provide for the application of timely corrective maintenance based on a machine's actual condition.

The DR ferrography report includes spectrochemical analysis, large- and small-particle quantity indexes, and the results from wear rate, intensity, and severity calculations. An analytical ferrography report includes specific type and quantity classifications of the metallic and nonmetallic debris present on the slide, a color photomicrograph of the ferrogram, an assessment of the sampled machine's overall wear status, and a detailed interpretation of the ferrography results.

Analytical Ferrography

Analytical ferrography uses a ferrograph fluid analyzer to concentrate on direct microscopic evaluation of the wear particles. A ferrogram slide is prepared by drawing the oil sample across a transparent glass or plastic plate in the presence of a strong magnetic field. Ferrous (iron) particles are attracted to the magnet and deposited onto the slide in decreasing size as the oil flows down the substrate. Nonferrous particles are deposited randomly, whereas ferrous particles line up in chains as a result of the magnetic flux. The result is a microscopic slide with the particles separated by size and composition. After deposition, the oil is washed away, leaving the particles clean, aligned with the magnetic field, and fixed to the plate. An experienced evaluator then examines the ferrogram to determine the composition and sources of the particles and the type of wear present.

Analytical ferrography is used to separate solid contamination and wear debris from a lubricant for microscopic evaluation. As stated earlier, spectroscopy is not able to measure wear particles larger than 7 μm in size. Although particle counting, ferrous wear concentration, and DR ferrography are able to detect the presence of larger particles, they cannot qualify their composition or origin. Analytical ferrography is able to identify wear particles, their composition, and their origin by visually analyzing them microscopically.

Microscopic examination of the debris reveals information about the condition of the equipment. Observing the concentration, size, shape, composition, and condition of the particles indicates where and how they were generated. Particles are categorized based on these characteristics, and conclusions can be drawn regarding the wear rate and health of the machine.

The composition of the particles can be identified by color. Heat-treating the slide causes specific color changes to occur in various types of metals and alloys. The particle's composition indicates its source. The particle's shape reveals how it was generated. Abrasion, adhesion, fatigue, and sliding and rolling contact wear modes each generate a characteristic particle type in terms of its shape and surface condition.

Solid contaminants also can be identified visually, provided that they are of a commonly found origin. Sand and dirt, fibers, oxidation products, rust, and metal oxides are examples of contamination debris that can be identified.

This test is qualitative, which means that it relies on the skill and knowledge of the ferrographic analyst. While this can have definite advantages, the interpretation is somewhat subjective and requires detailed knowledge of wear debris failure modes. The test procedure is also lengthy and requires the skill of a trained ferrographic analyst.

Elemental Spectroscopy

Elemental spectroscopy is a test that can determine the concentrations of 15–25 different elements ranging from wear metals and contaminants to oil additives. A spectrometer is used to measure the levels of specific chemical elements present in an oil. Two types of spectrometers are commonly used. Arc-emission spectrometers apply energy in the form of an electric arc to the sample. The other common type of spectrometer is the inductively coupled plasma (ICP) spectrometer. This operates on a similar principle, except that the energy is applied to the sample by a plasma flame rather than an electric arc. Spectroscopy is not able to measure solid particles larger then roughly 7 μm, which leaves this test blind to larger solid particles.

Elemental analysis works on the principle of atomic emission spectroscopy (AES). In AES, individual atoms within the sample, for example, iron atoms from wear debris, zinc atoms from a ZDDP-additive molecule, or silicon from silica (dirt) contamination are excited using a high-energy source. The atoms absorb energy from the excitation source and are transformed into a high-energy electronic state. Because of the laws of quantum physics, atoms do not like being in these excited states and rapidly lose the energy they have gained, mainly by emitting light energy. The energy of light emitted, which is inversely proportional to the wavelength, depends on the electronic structure of the atom and thus is different for each type of atom. So, by measuring the amount of light emitted at the characteristic emission wavelength for atoms such as iron, copper, zinc, and sodium, the concentration of each atom can be determined.

Significance

Monitoring the concentration of metallic elements can provide important information about machine and lubricant condition. By monitoring wear metals such as iron, copper, tin, and lead, rates of wear can be observed, and abnormal wear modes can be detected. Many contaminants have metallic components that can be monitored as well. Increases in contaminant metals such as silicon, aluminum, and potassium can indicate ingression of dirt, coolant, or process contaminants. Some additives also have metallic components. Monitoring additive metals can help indicate when a system has been topped off with the incorrect lubricant.

TAKING ACTION WITH OIL ANALYSIS

Reliability is enhanced when particles are controlled. The goal should be to lower contamination to an acceptable level. Therefore, it is imperative that action be taken to control contamination. There are three steps to achieving this goal:

Step 1: Set Oil Cleanliness Targets.   The proper cleanliness level is difficult to state in general. Where availability and reliability are of great importance, the oil cleanliness target should be higher. Water is a second key parameter to monitor and act on. The American Gear Manufacturers Association (AGMA)/American Wind Energy Association (AWEA) standard also includes guidelines for moisture contamination . The caution level is 0.05 percent (500 ppm), and the critical level is 0.10 percent (1,000 ppm). So an effective contamination-control program should aim for 0.05 percent or lower.

Step 2: Take Action to Reach Targets.   Two specific actions are required. First, reduce contaminant ingression. In other words, keep particles from entering the machine. This requires good housekeeping procedures in the storage, handling, and dispensing of oil. Ensure that the oil is kept clean and dry. Do not mix oils of unknown origin. Avoid cross-contamination by clearly labeling containers with the oil type.

New oil always should be introduced into the machine by means of a sufficiently fine filter (i.e., 3 μm). Studies show that new oil is often highly contaminated. Therefore, use offline filters and filter carts to clean and dispense new oil from drums and totes. New oil should be considered contaminated until the opposite is proved. Portable containers can be filled directly from the cart. During maintenance events, take great care to minimize the entry of contaminants. Add oil with the filter cart using quick-disconnect fittings. Breathers should have a filter and desiccant to remove ambient dirt and moisture; use labyrinth seals and V-rings. Next, improve filtration. Remove particles and water quickly. A well-designed filtration system will effectively remove not only solid particles but also moisture.

Step 3: Monitor and Maintain Oil Cleanliness.   Oil analysis will provide continual feedback on the condition of the machine and lubricant. It will also verify whether or not cleanliness goals are being met. If not, take the actions in step 2.

Determining If Oil Analysis Is Warranted

Determining whether a piece of equipment would benefit from oil analysis requires review of the following areas:

1. **Criticality of the equipment.** Typically, based on criticality ranking of the equipment, a decision on whether oil analysis is required is taken.
2. **Value of equipment/down time.** The value of equipment, in terms of replacement cost and/or operational loss while the equipment cannot be operated, should be considered when determining the value of an oil analysis program.
3. **Cost of analysis versus cost to replace oil.** Performing oil analysis on a small reservoir may not be cost-effective in all situations, but replacing the oil in a 100-gallon reservoir costs significantly more than oil analysis.
4. **Customer concerns.** The peace of mind that comes from knowing whether a problem exists or not is many times worth the investment, even with smaller reservoirs.

Oil Analysis Intervals

When should oil analysis be performed? The answer depends on the equipment, the frequency at which it is used, and service severity. Typically, oil analysis is performed at prescribed intervals. This interval schedule allows a baseline to be established to assist analyzers in isolating equipment failures. Oil analysis baselines establish typical values for wear metals and fluid characteristics for a particular piece of equipment. As more data points are collected, a more accurate baseline is established, and deviations from the norm are more easily distinguished. The suggested intervals shown in Table 11.2 are for conventional (nonsynthetic) oils only.

TABLE 11.2   Suggested Oil Analysis Intervals

Component	Interval
Hydraulic system	Monthly
Gas turbine	Monthly
Steam turbine	Quarterly
Gas compressor	Quarterly

Required Information for Proper Analysis

A great deal of analytical data are generated when oil analysis is performed. Proper interpretation of those data requires that complete and accurate information is provided by the customer to ensure that proper recommendations are made. If accurate or complete information is not provided to the laboratory, vital clues will be missed, and subsequently, interpretations and recommendations will not reflect this critical information. Laboratories generally provide a submission form to be completed by the customer and returned with the fluid sample. Different information may be requested by different oil analysis laboratories, but most labs will require the following:

1. Proper identification of the equipment/component: name, make, model, and year manufactured
2. Proper identification of the lubricant being used: brand name, type of product, viscosity grade
3. Time accumulated on the equipment since new or rebuilt
4. Date the oil sample was taken
5. Date the fluid and fluid filters were last changed
6. Indication that prior analysis has been conducted for the equipment (to allow for accurate interpretation)
7. Indication that recent maintenance has been conducted on the equipment

Analysis Results and Interpretation

Element analysis is important to determine acceptable concentration levels. Contaminant levels should always be zero in new fluids but may show up as a few parts per million based on cleanliness of the sump/reservoir or test equipment being used. Additives generally should correspond to the levels found in the new or unused sample of oil being tested.

Acceptable wear levels for wear elements can vary for the following reasons:

1. Identical equipment can create different wear rates.
2. How equipment is used affects wear rates (part load, frequent startup and shutdown).
3. The age of the equipment affects wear rates: internal wear is significantly higher during equipment break-in.
4. Length of time the oil has been in service affects wear rates.
5. Lubricant quality affects wear rates.

It is important to know the equipment's age when determining whether or not wear metal levels have become excessive.

The rate of wear is high while new components are breaking in or seating themselves. Once this has occurred, the rate of wear stabilizes and remains stable over the majority of the equipment's life. As equipment nears the end of its life or in the event of a failure, the rate of wear increases significantly. In order to properly interpret oil analysis data, it is important to know where a piece of equipment falls on this curve at the time the oil sample was taken.

Knowing the length of time a fluid has been in service is also critical to properly interpreting the level of a wear element.

Trending

Trending or trend analysis is the most accurate way to assess the condition of an oil or machine and involves a data set of three or more variables. Variables represent the instances the oil sample was taken, and in order to establish a trend, oil would have to be sampled at least three times. Although manufacturers publish general guidelines for wear rates and condemnation limits, the guidelines can be ambiguous or left to interpretation. Trending provides the most accurate method for determining normal values for interpreting oil analysis results and making recommendations.

Trending allows analyzers to look for abrupt changes over time. At least three samples must be on record for a trend to be determined. In the case where there is an inadequate amount of data, wear rates provide a guideline. For machines with a trend report, significant changes within the report will aid in determining potential problem areas.

ADVANTAGES OF OIL ANALYSIS

Reduce Unneeded Maintenance and Oil Waste

Oil analysis has frequently shown that oil is suitable for use beyond the general guidelines, helping to reduce oil waste and minimize both oil installation costs and routine maintenance. Oil analysis can be used to verify the extended drain interval performance of oils. Because of this, convincing plant managers to try extended drain intervals can be challenging. Oil analysis can be the tool that provides proof of safe operating conditions and improved performance.

Increase Equipment Reliability

Unexpected breakdowns cost more than scheduled maintenance. By taking advantage of oil analysis, owners can get a clear picture of the equipment's operating state, whether engine components are wearing at an acceptable rate or are showing signs of accelerated deterioration. By using this information, appropriate measures can be taken to repair troublesome components before an unexpected and costly failure occurs.

Normalize Equipment Operation with Trending Data

Developing wear trends that are characteristic of a particular piece of equipment permits precise recommendations based on the assessment of the lubricant and equipment.

Enhance Equipment Resale Value by Exhibiting Component Integrity

Oil analysis records can be used to show potential buyers that a piece of equipment is in good condition. If mechanical repairs are necessary, past repair records can corroborate that proper maintenance measures were taken.

Avoid Oil Mix-Up

Oil mix-up is one of the most common lubrication problems contributing to machinery failure. Putting the right lubricating oil in the equipment is one of the simplest tasks to improve equipment reliability. Checking the viscosity, brand, and grade of incoming new oil and checking any contamination of alien fluids help to reduce the chances of oil mix-up and keep the machine operating.

Contamination Control

Solid contamination (sand and dirt) accelerates the generation of abrasive wear. Liquid contamination such as moisture in oil accelerates machine corrosion. Fuel or coolant dilution in engine oil will decrease the viscosity, therefore generating more adhesive wear (rubbing wear). It is critical to keep the lubricating oil clean and dry all the times. This requires that you set cleanliness limits and continue monitoring contamination during machine operation.

FAILURE ANALYSIS

A comprehensive oil analysis suite may include tests such as ferrography and scanning electron microscopy (SEM)/energy dispersive x-ray spectroscopy (EDX), which are both time-consuming and expensive. However, these tests provide detailed and definitive information about machinery wear, such as what the wear particles are made of, where they come from, and how severe they are. Such information provides reliability professionals with information on a past or imminent failure.

CONCLUSION

Oil analysis is an integral part of the maintenance plan for manufacturing plants. Any piece of equipment that has a lubricating system is a potential candidate for oil analysis. A successful oil analysis program requires an organized and sustained effort. Lean manufacturing or Six Sigma initiative can not reach its goal without the processes to sustain improvement. Both the user and the laboratory must work closely together to lay the groundwork for the program and achieve the desired results.

ICML Questions

1. Kinematic viscosity is measured in
   a. centistokes (cSt).
   b. dunes.
   c. pascals.
   d. meters.

2. An oil's viscosity increases because of
   a. oxidation, degradation, or contamination of oil.
   b. extended oil drain intervals.
   c. the presence of water or another oxidation catalyst.
   d. All of the above

3. A decrease in an oil's viscosity is most commonly due to
   a. contamination with fuel or a solvent.
   b. extended oil drain intervals.
   c. oxidation.
   d. the presence of water.

4. Acid number (AN) is an indicator of
   a. total amount of acidic material present in the lubricant.
   b. acid buildup in oils as a result of depletion of antioxidants.
   c. acidic by-products formed from oxidation.
   d. All of the above

5. AN is measured by titration using
   a. ASTM D 664.
   b. ASTM D 666.
   c. ASTM D 668.
   d. ASTM D 669.

6. Abnormal decreases in base number indicate
   a. a depleted additive package.
   b. acid buildup in oils.
   c. oxidation of lubricant.
   d. All of the above

7. In most lubricating systems, water should not exceed
   a. 500 ppm.
   b. 1,000 ppm.
   c. 1,200 ppm.
   d. 800 ppm.

8. To quantify the amount of water in oil, which of the following tests is conducted?
   a. Simple crackle test
   b. Karl Fischer titration test
   c. Oxidation test
   d. Hygroscopic test

9. Demulsibility in lubricating oils is of utmost importance when the equipment is operating in
   a. humid climates.
   b. plants with water-intensive processes.
   c. the pulp and paper industry.
   d. All of the above

10. The demulsibility assessment test is conducted based on
    a. ASTM D 1401-02.
    b. DIN ISO 6614.
    c. ASTM D 1403-04.
    d. a and b

11. Elemental spectroscopy is a test that monitors
    a. lubricant condition.
    b. wear debris.
    c. contaminants.
    d. All of the above

12. In an ICP spectrometer, the energy is applied to the sample
    a. by a plasma flame.
    b. by an electric arc.
    c. by an electric spark.
    d. by an electron beam.

13. Spectroscopy is not able to measure solid particles larger than
   a. roughly 4 μm,
   b. roughly 6 μm.
   c. roughly 7 μm.
   d. roughly 3 μm.

14. The difference between DR ferrography, analytical ferrography, and spectroscopy is that
   a. analytical ferrography is able to identify wear particle composition and origin.
   b. DR ferrography measures wear particles larger than 7 μm in size but cannot qualify their composition or origin.
   c. spectroscopy is not able to measure wear particles larger than 7 μm in size.
   d. All of the above

CHAPTER 12

Creating an Effective Lubrication Program

Lubrication has a major impact on the reliability and life of equipment and the subsequent cost of maintenance and operational uptime. Although lubrication accounts for a very small portion of the maintenance budget, there is considerable industry evidence to suggest that a high percentage of rotating equipment failures is attributable to poor lubrication management. Investigation revealed that approximately 50 percent of premature bearing failures are due to issues such as too much or too little lubricant, lubricant contamination or cross-contamination with incompatible lubricants, lubricant chemical degradation, and use of the wrong type or grade of lubricant. Hence, to be at best-in-class level, companies should focus at great length on the management of their lubrication activities. Many companies try to achieve reliability of their plant and equipment by focusing on people and technologies. However, they put less emphasis on the lubrication component of the strategy, and consequently, their efforts do not always meet expectations.

Most people believe that just maintaining or adding lubricant to rotating equipment provides effective lubrication. There are many different methods to achieve the goals of your lubrication management program. This chapter discusses how to design a lubrication program to improve overall plant reliability.

LUBRICATION PROGRAM MANAGEMENT

The goal of every lubrication program should be to ensure that all equipment receives and maintains the proper levels of lubrication such that no equipment ever fails from inadequate or improper lubrication. For this to happen, we must follow the six R's of lubrication—right lubricant, right condition, right location, right amount, right frequency, and right procedures—for each piece of equipment.

Right Type

Two major factors in the selection of an oil-based lubricant are the correct viscosity and additives in the formulation. To get the right type of lubricant, you refer to the OEM manual and contact the OEM if you have any questions. With old equipment, the manual may be outdated, and better lubricants may be available. When in doubt, use your lubricant supplier along with the OEM manual. Most manuals are written for ideal conditions, but these guidelines do not address the real environment in which the equipment operates. To determine which type of lubricant is best for an application, one must understand the current situation. Look at application/environmental factors such as speed, temperature, load, vibration, moisture, and dust. Consider that:

- Temperature determines lubricant base oil type.
- Speed determines the required viscosity (at operating temperature).
- Load, vibration, and moisture determine the additive package.

As temperatures or speeds increase, the viscosity of a lubricant will decrease. *Viscosity*, the measure of a fluid's resistance to flow, is essential to protect equipment. You need a formulation that will provide a sufficient film layer of lubrication to reduce friction.

When selecting the appropriate industrial lubricant, keep in mind the application environment. For example, if you expect that a bearing will be subjected to significant amounts of water contact, then choose a lubricant with low water washout and high corrosion resistance properties. In situations where the bearings are operated at low speeds and extreme pressure, you can increase equipment reliability by choosing a higher base oil viscosity lubricant with high load capabilities. Selecting the right lubricant in each scenario can have considerable payoff.

Choosing the right lubricant is even more essential if you are operating in an aggressive environment, where there is typically high moisture and corrosion and extreme temperatures. It is even more critical to equipment life to select a lubricant that will maintain a sufficient film of lubrication to reduce friction, resist load and wear, and prevent corrosion.

Right Quality

Once the right type of lubricant has been selected, it is important to select a high-quality lubricant. Quality is both the ability of the lubricant to meet OEM specifications, based on performance on ASTM tests, and the cleanliness of the fluid in which it is delivered. You can have the highest-quality lubricant, but if it is not handled properly during delivery or storage, it will not perform the way you expect it to.

Right Amount

Too much lubricant in a system can be as destructive as too little, which is also a major failure mode. According to the American Bearing Manufacturers Association (ABMA), improper or insufficient lubrication is the cause of 64 percent of bearing failures. It is important to understand the various parameters surrounding the operation of any given bearing to properly select relubrication intervals. Overgreasing will lead to increased operating temperatures, resulting in energy losses and eventual bearing failure. Similarly, using too little grease will not allow the grease to properly carry the load applied to it, which will also result in bearing failure.

In splash-lubricated systems, enough oil needs to be splashed up for cooling and lubrication. Too high an oil level will cause churning, which overheats the oil; too low a level will not provide proper oil cooling and lubrication for bearings and gear teeth. Spur helical, bevel, and spiral bevel gears are lubricated with the gears dipping into the oil at twice the tooth depth. The OEM will provide information on the correct oil level.

Right Place

Once we have selected the right type of lubricant and the correct quantity to add, we need to apply it at the proper location. Adding the wrong oil to a lubrication point is not uncommon. This situation will usually go undetected until a problem occurs. With an oil analysis program, early-stage detection is more likely, thus helping to avoid possible equipment damage. All lubrication points should be properly labeled as to the lubricant to be added. Lubricant manufacturers provide lube tags for proper identification of the proper lubricant to be used at the lube point.

It is a good practice to use separate containers for different lubricant types. Mixing lubricants with different additive packages is not recommended. Normally, each lubricant supplier color-codes its tags by lubricant types.

Right Time

Once we have established our program with the right type, quality, amount, and place, we need to establish proper lubrication intervals. This timeframe is known as the *regreasing frequency*. This calculation is more complex. The machine's operating conditions must be collected along with some additional bearing information to derive regreasing intervals.

Temperature is known to affect both grease and oil, so naturally one of the first correcting factors that must be collected is operating temperature. The hotter the environment, the more frequently the grease must be replenished. The ambient contamination (how dusty the environment is) and moisture (how humid the environment is) work in a similar manner. The dirtier and wetter the environment, the more often the bearing must be regreased.

Simple things such as the bearing's physical position and vibration will also affect the grease's run-out and frequency of reapplication. If the bearing is mounted on a vertical shaft, the grease has a tendency to run out of the bearing more quickly, thus needing to be replenished more often. When the bearing is subjected to vibration, it causes the oil in the grease to separate from the thickener, allowing it to drain away from where it is needed much faster. Therefore, these two correcting factors—position and vibration—should be taken into consideration.

Right Lubrication Procedures

Once the correct lubrication type, quality, amount, and place and time have been determined, procedures should be put in place to maintain a lubrication program. This will ensure that the proper lubrication procedures for each piece of equipment throughout the plant will be followed by all maintenance personnel. Putting together a lubrication plan should become part of the maintenance standard operating procedures.

These factors should include:

- Reviewing storage and handling conditions
- Maintaining records of the correct lubricant type for each application
- Determining the proper amount of lubrication per day and frequency of relubrication
- Tracking mean time between failures (MTBR)/mean time between maintenance (MTBM)

By maintaining careful records, you can see patterns over time in performance and adjust accordingly.

HOW TO ACHIEVE THE SIX R'S

To achieve the basic six rights of lubrication, the following programs must be addressed so that it will meet your plant reliability goals:

- Lubrication program development
- Lubrication program implementation
- Lubrication program improvement

Lubrication Program Development Phase

In this phase, you will need to accomplish all the following tasks.

Develop an Equipment List.   The purpose of creating an equipment list is to develop a pre-liminary list of equipment that will be included in the lubrication program. Before a plant can begin implementing a lubrication program, it is necessary to create or obtain a current list of all equipment that requires lubrication. This list should include all types of equipment requiring lubrication, not just the usual pumps, motors, and compressors. Resources for this information include you're a computerized maintenance management system (CMMS), plant maintenance files, piping and instrument diagrams, and also a physical survey of the plant. At a minimum, the main output from this process should be equipment identification name, number, and process description.

Conduct a Lubrication Survey.   The lubrication survey will consist of a detailed lubrication inspection of all plant equipment. Each machine must be studied and its related characteristics recorded. Obtaining this information is time-consuming, and it may take several days or weeks to complete a plant survey. The lubrication survey is the only way of obtaining an accurate picture of current lubrication practices. It is also the basis on which future steps in improving the lubrication program are taken. You start with a list of the ideal oil analysis deliverables (answers to your important questions) and work backward. A question set on different aspect of lubrication reliability is developed to assess the present condition of the lubrication program.

Questions Relating to Lubricant Quality, Quantity, and Type

- Does the equipment have a tag identifying the proper lubricant type and amount to apply?
- A bearing nameplate at the site is extremely helpful. This helps in determining bearing size and relubrication volume. It is a good idea to have it documented. A plate at the bearing site stating the volume of grease required is helpful.
- Is the specified lubricant currently being used?
- What is the condition and quality of new lubricant deliveries?

Questions Relating to Storage and Handling

- Ideally, lubricants should be stored in an environmentally controlled area. All make-up oil containers must be kept sealed in a relatively clean environment. Are oil drum and make-up containers stored, in a designated area?
- Is everything clearly labeled?
- Is a lubricant in a storage container clean and fit for service?
- Does a color-coded grease cap exist identifying the required grease type?
- Are the grease guns color-coded and the volume per stroke clearly marked on the gun?
- Have two incompatible greases or lubricants been mixed?

Questions Relating to Sampling

- The primary sampling location must be clearly marked to ensure the correct location is used. Sampling ports should be labeled and prepared, and samples should be taken in a reliable and consistent external contaminant-free manner. Are there any written guidelines and procedures?
- Is a sampling valve installed in a primary location?

Questions Relating to In-Service Lubricant Health/Condition

- Have any vital additives become depleted and, if so, how much?
- Has the base stock been impaired by thermal degradation, oxidation, or hydrolysis?
- Has interfacial tension changed, affecting foaming, air release, and demulsibility properties?
- Has the oil's viscosity changed as a result of evaporation, shear, radiation, contamination, or various chemical reactions?
- What is the residual life of the lubricant?
- Is a lubricant in a storage container fit for service?

Questions Relating to Lubricant Contamination

- How is the general condition of the reservoir (rusted, corroded, leaking)?
- It is extremely important to remove moisture and contaminants transported via the exchange of air during ventilation? Open vents are generally not acceptable. Are there any open vents?
- Are there any leaks in the system?
- Has the lubricant been mixed with other fluids or lubricants?
- Has the lubricant become contaminated with soot, dirt, fuel, water, process chemicals, or coolant?
- Does the environment contaminate the oil (steam, dust, dirt)?
- Is there evidence of abnormal operating temperatures, pressures, or duty cycle?
- Are the test results tracked and examined for trends?
- Are procedures and analyses defined?

Questions Relating to Training Needs

- Find a technician. Pick a work center. and ask how the mechanic determines which type of grease to apply. How does the mechanic complete the lube process?

Selecting the Right Lubricant

The OEM manual is always the first place to go to identify lubricant selection requirements. Second, engaging the vendor lubricant representative is absolutely crucial. The vendor you choose should be able to cross-reference the OEM with its available lubricants, oil, and greases and select the appropriate one for your machinery. When an organization lacks lubrication expertise, the vendor can be an excellent source of support.

Consolidation of Lubricants (If Applicable)

Once lubricants have been selected for each piece of equipment in the program, it is important to review the list and determine whether there are any opportunities to reduce the total number of lubricants that will be used. Aim to rationalize selected lubrication products to one supplier as far as practical. Ideally, this can be achieved by inviting a representative from the chosen supplier to visit the site and make recommendations.

The benefits from standardization and reducing the number of lubricants have the following effect on the program:

- They reduce the number of lubricants that have to be purchased.
- They reduce the number of lubricants that have to be documented.
- They save on transport costs.
- They reduce the risk of cross-contamination.
- They ease the management of procurement and storage of lubricants.

Lubrication Manual Development

The rationale behind creating a lubrication manual is to have all pertinent lubrication information gathered in one place. After all the time and effort expended to locate and collect the data, it is worthwhile to consolidate that information into an electronic lubrication manual so that it can be easily referenced by all plant personnel over time. In addition, by having an electronic lubrication manual, your program can become a living organism. At a minimum, a detailed lubrication manual should include the following:

- Equipment number and description
- Lubricant section from the OEM manual
- Selected lubricant technical data sheet
- Selected lubricant Material Safety Data Sheets (MSDSs)

ORGANIZATION AND PLANNING

The actual act of lubricating, filling, or visually monitoring a specific asset within a manufacturing facility requires planning and scheduling. The plan should be mapped out so as to define the points, lubricants required, frequency and quantity needed for relubrication, refilling or changing the lubricant, as well as the personnel and procedures required to perform and document the specific lubrication task. To streamline this function, we need to have an organization and planning for lubrication function.

The lubrication organization should be centralized to obtain the benefits gained by rationalizing and then standardizing lubrication-related items. Therefore, the four main stakeholders or role players will be the maintenance department (customer), the planning department (which will distribute the schedules), the lubrication services team (which will execute the tasks), and the lubrication subject-matter experts (who must provide up-to-date recommendations and best practices regarding plant lubrication needs). These experts could include plant lubrication engineers, lubricant suppliers, OEMs, and/or other lubrication engineering subject-matter experts.

LEADING THE PROGRAM

The key to this program's success is selecting the right person to bring the major role players together for a common goal. The leader is the hub of the program and must also have the authority to implement changes and manage them accordingly. It is advised that a central department such as a maintenance planning department lead the process and manage the functions once the program is in place. This is the best scenario because the program procedures need to be implemented throughout the entire plant. The main objective will be to manage lubricant usage, prevent contamination, and minimize adverse environmental impacts associated with lubrication activities. These tasks will generate interaction among the four players. For example, maintenance will blame the supplier for breakdowns. On closer inspection, contamination (gland leaking) could be found to be the root cause of equipment failure. As a result, the supplier will request maintenance to remove the contaminant (repair the leaking gland) to protect itself. To eliminate waste, the supplier will continue to request that maintenance repair oil leaks. This interaction can become cumbersome. The roles and responsibilities of each player should be clearly defined to prevent miscommunication and misunderstanding of the concept.

Role of Lubrication Subject-Matter Experts

- Provide lubrication survey
- Ensure minimum number of lubricants to optimize performance

- Continuously improve lubrication standards
- Recommend better maintenance practices

Role of Planning Department

- Schedule survey of maintenance program

Role of Maintenance Department (Customer)

- Repair oil leaks
- Make lubricating points accessible
- Avoid/remove contaminants from equipment

Role of Staff Performing Lubrication Functions

- Ensure clean handling of lubricants
- Lubricate according to schedule
- Report malfunctions such as oil leaks, inaccessible lubrication points, and contamination of equipment

BUILDING THE LUBRICATION PREVENTIVE MAINTENANCE (PM) PROGRAM

Set Lubrication (PM) Frequency

Before entering the lubrication task into the computerized maintenance management system (CMMS) or enterprise asset management system (EAMS), it is necessary to determine the PM-related lubrication data. The key questions to ask are what kind, how much, and how often? Answering these questions will start your lubrication PM program. This information, along with the other data collected, will be put into the system and used to generate the service schedule. Some basic steps can be followed to achieve PM program success:

- The OEM manual is an important source of information for building your PM schedules. The lubrication section of the manual should describe the locations on each machine to be lubricated, the type of lubricant to use at each location, the quantity of lubricant to be applied, and how often to apply the lubricant.
- Schedule the lubrication routines in a CMMS. Regardless of the system used, the maintenance department's daily lubrication activities should be organized. Responsibility should be assigned for each lubrication function.

- Color-code and symbol-code all lubricants and lubrication points. This will minimize mistakes when delivering the PM routines.
- Change filters when indicating devices denote plugging or when fluid analysis reveals that a change is needed.
- Keep maintenance records on each system as an aid in determining good PM techniques. Records should be audited occasionally to ensure accuracy and thoroughness.

Logistics and Supply Chain

When selecting suppliers to augment lubrication PM program implementation, consider "partners" capable of providing a full range of goods and services, including lubricants, lubricating systems, lubricant analysis, failure analysis, and services that may need to be outsourced because of personnel constraints or specialization not found within the facility. Their help may be required for lubrication planning and scheduling design, lubricant analysis program design, standard procedures generation, storage room design and training, and lubrication tools and automatic lubrication system recommendations.

Necessary Lubrication Equipment Procurement

When purchasing the necessary equipment for your lubrication program, you must remember that you want the lubricants in the right condition. Properly lubricating equipment involves the use of equipment to both store and apply lubricants at set intervals and as needed. In storing lubricants, you want to ensure that the lubricants are used in a first in, first out (FIFO) manner. Remember that all lubricants have a shelf life. When applying lubricants, the following equipment is recommended (but not limited to):

- Storage racks
- Sealed plastic heavy-duty oil-dispensing containers
- Grease guns
- Bulk lubricant storage containers
- Filter carts

Implementing the Lubrication Program

The second phase in building a lubrication program is the implementation phase. Once all the data have been collected and gathered in the development phase, the information must go somewhere. If the information is not already in the plant's CMMS or EAMs, it must be input into it. Once in the CMMS or EAMs, you must ensure that the PM and task frequencies are set. Once the frequencies have been set, create the lube routes. Review the routes for clarity and consistency. If any changes need to be made, ensure that they are made, and then set the inspection schedule (Figure 12.1).

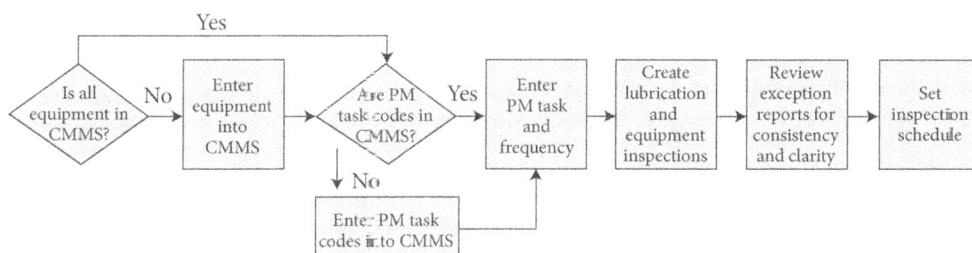

FIGURE 12.1   Lubrication implementation flowchart.

Another key factor in implementing a program is to have your safety practices developed and in place. When it comes to lubrication safety, there are a number of unique aspects regarding the use and handling of lubricants. Because they are designed to minimize friction in machines, lubricants are slippery. When a lubricant is spilled or leaks onto the operating floor or any other undesirable location, it can lead to a high-risk situation that must be immediately attended to in order to prevent personal injury.

In addition, because they are for the most part hydrocarbon derivatives, lubricants are flammable. The proper fire-hazard precautions must be taken. Finally, some lubricants can cause personnel health problems when the lubricant comes into contact with skin. Items for consideration for your safety practices include, but are not limited to, the following:

- Material Safety Data Sheets (MSDS) must be available and reviewed.
- Leaks are under control.
- Spill response measures are in place.
- Handling practices maintain a safe environment.
- The operation of lubrication equipment use is understood.
- Sampling procedures are in place and followed.

In the lubrication implementation phase, lubrication storage and handling and contamination control play important roles.

Lubricant Storage and Handling

It is recommended that the lubricant storage area be a safe and controlled environment that complies with health and safety regulations. Required documentation must be available and accessible. The area should be equipped with spillage- and fire-control devices. Many facility operators are unaware of the dangers improper lubricant storage and handling practices create and what inevitable fate they can lead to in terms of equipment reliability and lifecycles. This is one area on which even the best companies fail to focus. Here are some tips for proper lubrication storage and handling:

- **Design of the lubrication storage areas.** The purpose is to ensure ease of use and limit handling mistakes. The storage areas should be well lit and organized, and they should include provisions for bulk storage. An area also should be set aside for top-off containers and grease guns.
- **Quality control program.** This is to ensure that the correct oils and lubricants are being delivered and that the cleanliness of the delivered lubricants is up to current target particle and moisture cleanliness levels. Checks should also be made to ensure that oils have not exceeded their recommended shelf life. Finally, the quality control program should routinely sample and test oils for contamination.
- **Labeling and identification.** The labeling system can efficiently ensure that the right lubricant is used at the right location and prevent cross-contamination. The labeling system should be simple and easy to implement.

Another important task is to determine whether your dispensing tools and equipment help to minimize the ingress of contaminants. Lubricants and tools should be properly identified to avoid cross-contamination.

Lubricant Dispensers and Grease Gun Storage

Storage for dispensing containers, grease guns, and rags is another important step to ensure that contaminants are not introduced into the lubricants as a result of poor housekeeping. These tools should have their own dedicated fireproof storage cabinets for easy access and organization.

Grease guns should be stored in a clean, dry, and controlled environment. They are precision tools that must be cared for accordingly. Regular cleaning and inspection of proper function are a must!

Oil dispensers should be of the sealed type, and special care should be taken when transferring oil from a bulk container to top off dispensers to avoid contamination. The use of portable lubricant dispensing carts is the preferred method for transferring large amounts of oil to machines with large gearboxes and reservoirs.

Contamination and Condition Control

Contamination in lubricants accounts for approximately 25 percent of all lube-related bearing failures. Only clean lubricants should be used on machines, so it is essential to know what procedures are in place to avoid the introduction of contaminants into new lubricant. How are sources of contamination identified and controlled? Are ISO cleanliness codes and water-level content defined for critical assets, and how often do you control these levels?

Cleanliness control centers ensure that the lubricant is stored safely, is clean, and is transferred in a contamination-free environment. If spilled, lubricants can contaminate the environment. Therefore, most maintenance strategies today work to prevent such contamination from occurring. This is also a key ingredient in world-class manufacturing standards.

Standard Operating Procedures (SOPs)

As in other disciplines (operating or repairing machinery), SOPs can help guarantee repeatability and quality work. In addition, SOPs assist in the training of lubricators and enable you to track and communicate equipment condition.

Improving Your Lubrication Program

After your lubrication program has been developed and implemented for a certain amount of time, you must review your equipment history. In order to improve your program, you must identify, from a detailed equipment history, where your lubrication-related issues occur within your facility. From the equipment history, check whether the mean time between failures (MTBF) is acceptable. If not, conduct a root cause failure analysis (RCFA) to see if any lubrication-related failures are hampering your reliability goals. Other areas of concerned in improving your lubrication program include both your lubricant analysis program and the goals and metrics used to track program progress.

Lubricant analysis is commonly used as a diagnostic tool in facilities. However, many oil analysis programs frequently lack the proper setup and utilization of data needed to gain maximum benefit. All too often an oil analysis result is provided to the customer in a hard-copy format with generic recommendations, and usually it is filed, never to be looked at until time for an audit. In such situations, oil analysis provides little or no value to the organization and to the overall reliability of the facility.

The establishment of goals and metrics is key to improving a lubrication program. The selection of specific program goals and the development of key performance indicators by which to measure the progress toward those goals both largely depend on the maturity of the program. Unfortunately, the development of goals and metrics continues to be an area of weakness in many lubrication programs. Although most organizations have established corporate and plant-specific goals and metrics aimed at overall operating and maintenance improvements, few programs have established goals and metrics at the technology level.

This is a key program element that is necessary to ensure lubrication program excellence. It is also important to have a clear understanding of the current status of the program, and it is equally important to have both vision and focus on the continued improvements that can be made to the program to realize effective and efficient fulfillment of the lubrication needs of the organization.

Performance Measurements

To be assured that your lubrication program is effective requires the measuring of selected parameters and the continuous trending of the data over a period of time. One of the parameters may be as simple as the number of failures due to poor lubrication or a ratio of the number of lubrication-related failures per total failures. Another good measurement comes from planning and scheduling the lubrication activities. Calculate and record the schedule compliance ratio. This is the number of lubrication work orders completed as scheduled. Not only does this tell you that the work is getting done, but it also provides some history to monitor frequencies of application. It is suggested that you use all this information to get started and then back off if all goes well.

Continuous Improvement

Continuous improvement is an important element of a comprehensive lubrication program. It has often been stated that in order to get better, it is necessary to understand where you are. To sustain reliability of the plant, the lubrication program should be reviewed and improved on an ongoing basis. By using an appropriate audit or self-assessment process, an organization will have a roadmap to address and evaluate where it currently stands and where its focus needs to be. This can be done as an annual review led by the maintenance engineer and the lubrication technician. A thorough review of each lubrication schedule can be made every three years or some other predefined review cycle. Selected schedules can be staggered so that a third of the PM schedules are reviewed annually. Equipment history for the equipment reviewed should be evaluated. Failures occurring as a result of poor lubrication should be examined to determine the root cause.

In addition to annual reviews, event-based reviews can be made when equipment fails and lubrication is the suspected cause. The intent would be to examine lubrication frequency and methods to identify ways to optimize the lubrication routine. Finally, the lubrication program should be audited regularly to ensure that it is adhered to and that the goals are being achieved. By using an appropriate audit or self-assessment process, an organization will have a roadmap to address and evaluate where it presently stands and where its focus needs to be.

It must also be understood that continuous improvement is needed to create a living program that continually changes to ensure both equipment reliability and ultimate cost effectiveness.

The audit requires a detailed assessment of your practices and their fundamentals. A lubrication audit questionnaire is structured to cover the different aspects of the lubrication program. The results form the basis of the improvement plan.

- **Lubricant selection.** Have you undergone a detailed selection and consolidation process to optimize both the lubricants for your application and the inventory? Is it reviewed by the oil analyst.
- **Storage and handling.** Does your lubrication storage room provide a clean and safe environment for your lubricants and personnel?
- **Lubricant application.** How do you ensure that only the right lubricant is being supplied in the right way at the right point?
- **Lubricant analysis.** Do you have a formal lubricant analysis program? Does it assess lubricant health, machine health, and contaminant monitoring? Does it address the root-cause analysis?
- **Lubricant contamination and condition control.** How do you monitor, remove, and control lubricant contamination?
- **Lubrication program management and personnel development.** Do you have a structured and consistent process to execute and follow up on your lubrication tasks? Does it include key performance indicators, training, and constant improvement goals?
- **Lubrication practices standardization.** Are all your procedures properly documented, implemented, and updated?
- **Environmental, health, and safety (EHS) practices.** Do you consider EHS regulations in your lubrication program?
- **Automatic lubrication system (ALS) policies and practices.** Are you taking full advantage of available technologies to optimize your machinery lubrication conditions?

The output of the lubrication audit is a comprehensive report of your current lubrication program and its efficiency. It includes a detailed list of strengths and improvement opportunities along with a series of recommendations to guide you in taking your lubrication program to a world-class level.

Training and Education

Educating the workforce is a key first step in launching your lubrication program. The operators and mechanics who are directly responsible for lubrication must be suitably trained, and individuals who are indirectly involved in the lubrication program also should have at least a basic awareness of the program's goals, primary benefits, and procedural requirements. The training program may cover all aspect from the physical activity of applying lubricants to the effects of misapplication and proper audits. In addition, training should include development plans necessary to maintain, monitor, and improve the lubrication management program.

The final piece to implementing a lubrication program lies in training the individuals who will be performing the tasks at hand. All persons performing lubrication-related work must be properly trained to ensure that the elements of the lubrication program are effective and documented. If these documents exist, they must be reviewed and evaluated. When reviewing these documents and procedures, you must ensure that they support the maintenance strategy, address the purpose of the task, and give clear guidance and direction to complete all tasks in a safe, effective, and efficient manner.

CONCLUSION

Lubrication is the lifeblood of any machinery, and the lubrication program should be treated as a critical piece in your overall reliability process. By leveraging proper lubrication techniques, machine uptime may be extended, and maintenance and operating costs can be lowered, greatly improving the overall lifecycle of the equipment.

Whether you are developing a new program or reviewing an existing one, the lubrication audit should assist you in evaluating and improving your program. Paying attention to detail and following your program are keys to success and achieving program goals. Achieving program goals ultimately should lead to higher uptime, greater throughput, and lower operating costs, resulting in greater profitability.

ICML Questions

1. Which of the following does not belong to the six R's of lubrication?
   a. Right lubricant
   b. Right supervision
   c. Right amount
   d. Right frequency

2. The purpose of a lubrication survey is
   a. to get a picture of current lubrication practices.
   b. to determine current lubrication frequency.
   c. to document the source(s) of lubricant.
   d. to quantify the quantity of lubricant used.

3. Overgreasing of electric motors will
   a. reduce the temperature of bearings.
   b. lead to bearing failure.
   c. lead to long lives of bearings.
   d. be the same as normal greasing.

4. Which of the following is *not* a part of continuous improvement of a lubrication program?
   a. Annual reviews of the lubrication program
   b. The lubrication audit
   c. Training and education
   d. Follow-up of lubrication tasks

5. Which of the following is *not* a part of performance measurement in a lubrication program?
   a. Ratio of the number of lubrication-related failures per total failures
   b. Number of lubrication work orders completed versus scheduled
   c. Audit of storage and handling of lubricants
   d. Continuous trending of data

6. To establish safety practices for use of lubricants, which of the following points is *not* relevant?
   a. Leaks are under control.
   b. Spill response measures are in place.
   c. Handling practices maintain a safe environment.
   d. Root-cause analysis of lubrication failures is undertaken.

7. Which of the following is *not* a good PM practice?
   a. Using the OEM manual as an important source of information
   b. Keeping maintenance records on each system
   c. Selecting the lubricant supplier
   d. Changing filters when plugging is seen

Index

About the Author

Trinath Sahoo, Ph.D., is General Manager (Maintenance and Reliability) at M/S Indian Oil Corporation, Ltd. Dr. Sahoo earned his Ph.D. degree from The Indian institute of Technology (ISM), Dhanbad. He has 30 years of experience in various fields such as engineering design, project management, asset management, maintenance management, lubrication, and reliability.

He has published many papers in several different journals: *Hydrocarbon Processing, Chemical Engineering, Chemical Engineering Progress*, and *World Pumps*. Some of his articles were also featured and published as cover stories in magazines. Dr. Sahoo has also spoken at many international conferences on the topics of lubrication and reliability. Dr. Sahoo was the Program Leader for a reliability enhancement project for different refinery and petrochemical sites of M/S Indian Oil Corporation, Ltd.

Although plants spend billions of dollars annually on equipment reliability initiatives, many overlook lubrication as an area of opportunity worthy of renewed focus. Dr. Sahoo considers lubrication to be a foundational component of equipment reliability and a best practices program. He is the author of the bestselling book, *Process Plants: Shutdown and Turnaround Management*.

www.ingramcontent.com/pod-product-compliance
Lightning Source LLC
Chambersburg PA
CBHW082004190326
41458CB00010B/3063